KT-373-528

Epitaph for the Elm

Dedication
William Wilkinson aged 2
(elm aged 200)

GERALD WILKINSON

Epitaph for the Elm

HUTCHINSON OF LONDON

Consultant editor Richard Mabey
Art editor & designer Bernard Higton
Picture researcher Christine Vincent

Hutchinson & Co (Publishers) Ltd
3 Fitzroy Square, London W1P 6JD

London Sydney Melbourne Auckland
Wellington Johannesburg and agencies
throughout the world

First published 1978

Phototypeset in Monophoto Photina 747 by
Oliver Burridge Filmsetting Ltd, Crawley

Printed in Great Britain by
Sackville Press Billericay Ltd

Bound by
William Brendon Ltd, Tiptree, Essex

ISBN 0 09 131450 X

Contents

The elms are dying. Gaps appear in familiar lines of trees that we never bothered to think of as elms. Half the tall trees, and many smaller ones, in the roadside hedges seem to have been elms, we notice, now they are so unhappily conspicuous. Rows of majestic trees that stand across meadows – perhaps survivors of vanished hedges – are now patchily wintry in the middle of summer. Great elms, landmarks to nowhere in particular, are shattered from green to grey, and will soon be gone, never to be replaced in our lifetime. Rich corners of rustic England are one year a little yellowed, the next as bare as battlefields. Those dusty summer lanes in the heart of England that were half black shadow, half tattered sleepy sunlight among cobwebs, nettles and leafy elm shoots will soon be exposed to an unnatural glare – and on wet days will be as exposed as the fields.

Little woods here and there which seemed to be of no particular tree now sadly, pointedly, show what they were. Houses and cottages once sequestered among two or three friendly trees are now threatened by their dead shapes. Suburbs once leafy are now twiggy, and will become more so. Country church towers, over-topped by elms, look forward to unwonted prominence, and rooks will have to find new homes. By village greens, trees that have been large while four generations grew up and played cricket, or flirted, will be gone.

We took them for granted as an inseparable part of our characteristic lowland landscape: loved by the poets and tolerated by the farmers. But the elms were not always tall. How did they, or their parent trees, come to be there? This book is about their history and natural history, their place and meaning in the landscape, their usefulness to man.

Part 1
The Elm in the Landscape of the Mind

The Tree of Ill-omen

'He will wait for me under the elm', says a French proverb, meaning he will not be there – perhaps because *sous l'orme* would be a stupid place to wait. Country people seem convinced of the malevolence of elms, which drop heavy branches unexpectedly. No injuries have been reported, but the reputation stays:

Elm hateth
Man and waiteth

is quoted by H. L. Edlin (1949) as a quaint couplet – older, and certainly neater, than Kipling's lines

Ellum she hateth mankind, and waiteth
    Till every gust be laid,
To drop a limb on the head of him
    That anyway trusts her shade . . . .

People used to sleep under trees, and would choose an oak; even the drip of the leaves was thought to be beneficial. John Evelyn wrote that 'the very shade is wholesome' and that sleeping or lying under the oak was a remedy for 'paralytics'. There is no such praise for elms in English lore, but Pliny said their shade was so gentle and benign that it nourished whatever grew beneath it. The shade of most field elms is not dense: grass and nettles grow, and the cattle, at least, do not mind waiting under the elms.

One modern poet, Ted Walker, has summarized our ambivalence to elms.

Elms are bad, sinister trees.
Falling, one leaf too many,
They kill small boys in summer,
Tipped over by a crow's foot
Bored with the business of leaves.

An uneasiness attends
Dead elms – timber for coffins,
ammunition boxes . . . .

'And', he continues, 'breakwaters', which are the subject of his poem, and in which he finds beauty.

Elm logs on the cottage fire do nothing to improve the reputation of the tree. Unless they are really dry they are slow to burn. English people, always spoilt, when not actually starving, cannot cook inferior joints of meat and are impatient with slow-burning fuel: ash and holly will burn straight from the living tree.

Elmwood burns like churchyard mould
E'en the very flames are cold

runs an anonymous and undated verse extolling the virtues of apple wood and ash logs, wet or dry.

Another anonymous rhyme, ending in praise of ash, is fairer to elm. The chestnut mentioned is probably horse-chestnut as it gives little heat.

Oak logs will warm you well
If they're old and dry.
Larch logs of pinewood smell,
But the sparks will fly.
Beech logs for Christmas time;
Yew logs heat well;
'Scots' logs it is a crime
For anyone to sell.
Birch logs will burn too fast,
Chestnut scarce at all;
Hawthorn logs are good to last
If cut in the Fall.
Holly logs will burn like wax:
You should burn them green.
Elm logs like smouldering flax
No flame to be seen.
Pear logs and apple logs,
They will scent your room.
Cherry logs across the dogs
Smell like flowers in bloom.
But ash logs, all smooth and gray,
Burn them green or old;
Buy up all that comes your way;
They're worth their weight in gold.

Such verses are not as ancient as their language pretends and there is little wisdom in them. All the same, most people cannot store and dry a whole winter's supply of wood fuel. If you have a large fireplace you can burn new elm logs. They will smoulder all day, but not without heat, and will glow red by evening. There is a good deal of elm lop and top available these days, for the obvious reason, and enthusiasts for using natural resources might care to look into the matter of wood-burning stoves on the Norwegian model.

The Beauty of Elms

Contributing little or nothing to our mythology and getting a poor press in the folk lore did not prevent elm timber being at least the second most useful to the village carpenter. William Ellis, the author of *The Timber Tree Improved*, 1745, tells of 'A person in the parish of Ivinghoe, in Bucks', who 'planted an elm himself, and at sixty years old he cut it down, with a hundred foot in it; that he sold, at one shilling per foot' – which is a sort of poem, if you like.

For poets of the landscape, elms always contribute something definite: they have a lot of 'character', more than just tall trees, and their presence can often be felt even when their name is not mentioned. I imagine elms in:

Between the open door
And the trees two calves were wading in the pond,
Grazing the water here and there and thinking,
Sipping and thinking, both happily, neither long.

(Edward Thomas, *Up in the Wind*)

Thomas, who knows his trees and weaves their names beautifully into his verse, has a poem about a farmer who loved women, horses and trees:

For the life in them he loved most living things,
But a tree chiefly. All along the lane
He planted elms where now the stormcock sings
That travellers hear from the slow-climbing train.

Many years since, Bob Hayward died, and now
None passes there because the mist and the rain
Out of the elms have turned the lane to slough
And gloom, the name alone survives, Bob's lane.

Elsewhere, in *October*, he gives us vivid small pictures of elms from first-hand observation:

The green elm with the one great bough of gold
Lets leaves into the grass slip, one by one, –
The short hill grass, the mushrooms small, milk white,
Harebell and scabious and tormentil . . . .

And 'after five minutes of thunderstorm', on *May the Twenty-third*,

The elm seeds lay in the roads like hops . . . .

Only Edward Thomas makes sense of the gloom, the rainy days and the mess of the countryside. Elms are part of this confusion. Landscape to him is, as it should be, where he lives and breathes, and not a picture in a frame. He says in *The Green Roads*, and we must believe him,

. . . all things forget the forest
Excepting perhaps me . . . .

He was, it seems, too modest to think of writing poetry until he was near the end of his short life. He was born on 3 March 1878. We leave him, here, sitting by a woodpecker's hole in the crest of a blizzard-felled elm, conversing with a ploughman, intermittently, as the horses turn. One of the ploughman's mates has been killed in France, as Thomas himself was soon to be, in 1917.

Now if
He had stayed here we should have moved the
tree.
And I should not have sat here. Everything
Would have been different . . . .
The horses started and for the last time
I watched the clods crumble and topple over
After the ploughshare and the stumbling team.

Clods of earth, and themes of life and death, these are part of the modern imagery of the elm.

The word 'tree' is weak and inadequate for its meaning; 'elm' is at least heavy and bell-like. For John Betjeman, very much the poet of the twentieth-century landscape, elms are important, close to everything he loves most.

Betjeman, unlike most other poets, does not mind being particular about species. Miss J. Hunter Dunn's father's euonymus, for instance, slips into the metre, disconcertingly accurate – not a poet's plant, you would have thought, until then, even though the spindle tree of the countryside is so poetic. Betjeman loves proper names, of course, and his tree flora is wide-ranging. Poplars are trembly (in Wembley), willows in Essex are blue; ashes, in leathery Lambourne, are feathery. Pines murmur in Pinner, or are scented in Surrey; laburnums lean in suburban Middlesex. Familiar trees become strangely evocative in unfamiliar places: beeches rock and thunder in an Irish gale. Also in Ireland are 'twisted trees of small green apple', and

Winding leagues of flowering elder
Sycamores with ivy dressed.

In Betjeman's England the native oak appears rarely, and in a strange light: 'Stumps of old oak trees attempted to sprout,' in sinister Lincolnshire (when the pony fell dead). Limes, so familiar in our townscape, appear only once, I think – in *Cheltenham*. Slough, waiting for a merciful bomb, is treeless apparently.

But elms are everywhere, and always different. They are often dark Eumenides, as in his childhood in usually sunny Highgate, where there are also hornbeams, laurels and cedars – a nice mixture.

The elms that hid the houses of the great
Rustled with mystery. . . .

But his elms are unpredictable:

See the rich elms careering down the hill
Full billows rolling into Holloway . . . .

The playground nettles nod at a horrible crime in the classroom 'On Highgate Hill's thick elm-encrusted side'. Later on, 'Waiting elm boughs black against the blue' witness a chilly moment of adolescent excitement at Marlborough. They wait again at Avebury as desire deepens into passion: then

there was love
Too deep for words or touch. The golden downs
Looked over elm tops islanded in mist,
And short grass twinkled with blue butterflies.
Henceforward Marlborough shone.

At Oxford there are '. . . guardian elms before St John's'. From his rooms

The wind among the elms, the echoing stairs,
The quarters, chimed across the quiet quad. . . .

All this remained for someone else, as, in disappointment, he reflected on his destiny.

The lean acacia tree in Trinity
Stood strong and confident, outlasting me.

– no elm to witness that moment of personal defeat, which was in fact a victory for poetry:

Farewell to blue meadows we loved not enough
And elms in whose shadows were Glanville and
Clough. . . .

Into the wider Oxford landscape the trees are etched with truth and delicacy: it is May Day

And a Cherwell mist dissolveth on elm-
  discovering skies.

Then to a landscape wider still the bells of *Bristol* ring, in a verse composed entirely of the digits 1 to 5, and

. . . an undersong to branches dripping into pools
  and wells
Out of multitudes of elm trees over leagues of
  hills and dells
Was the mathematical pattern of a plain course
  on the bells.

In *Dear Old Village*, bell-depleted and motorbike-ridden,

The elm leaves patter like a summer shower
As *lin-lan-lone* pours through them from the
  tower.

The bells are silent perhaps in 'deepest Essex', where '. . . out of elm and sycamore/Rise flinty fifteenth-century towers'. In Hertfordshire the elms, in a strange and beautiful metaphor, are 'shadowy cliffs' with 'pale corn waves rippling to a shore'. They are dark, too, in Sussex, where they are the backdrop to scenes of devilish behaviour (followed by Post-Toasties and Golden Shred): 'Dark rise the Downs from darker looking elms. . . .' And those are black trees, surely, that witness one of the oddest of Betjeman's poetic images:

The heart of Thomas Hardy flew out of Stinsford
  churchyard
A little thumping fig, it rocketed over the elm
  trees.

In another funeral poem, *In Memory of Basil. . . ,* is another weird image of flight, and an unexpected adjective for elms:

On such a morning as this
  with the birds ricocheting their music
Out of the whelming elms
  to a copper beech's embrace
And a shifting sound of leaves
  from multitudinous branches. . . .

We turn from symbolism to nostalgia and less subtle rhythms in *Cheltenham*, where

. . . a summer wind
Was blowing the elm leaves dry,
And we were seventy-six for seven
And they had C. B. Fry.

In this poem it is the limes which have the last word, as they do, unfortunately, where, out of *The Town Clerk's Views*,

Burford's broad High Street is descending still
Stone-roofed and golden-walled her elmy hill
To meet the river Windrush. . . .

At least the limes of Burford will be green again next year, unlike the many elmy hills of the surrounding countryside which blurred the poet's vision.

Betjeman's elm-filled landscape is essentially fading, transient, nostalgic: it seems to have passed from us, leaving a highly detailed image, like an old photograph going brown at the edges. He seeks it in the colour-plate books of his childhood, as he explains in his poems more than once, or remembers it from experience which now belongs to another era. The landscape is not the less real for this; and it can still be found. England, always changing, changes at different speeds over its varied surface. Betjeman enables us to select, as we rush through the lanes at an unpoetic pace, our very movements sufficient to cause minor ecological shifts. Now who will be the poet of the motorways?

Betjemanland is seen from bicycles, branch-lines and trams, moving in classical metres. In *Parliament Hill Fields*,

  mighty elms retire
Either side of Brookfield Mansions

as the tram wanders on through the gloomy hills of London, NW.

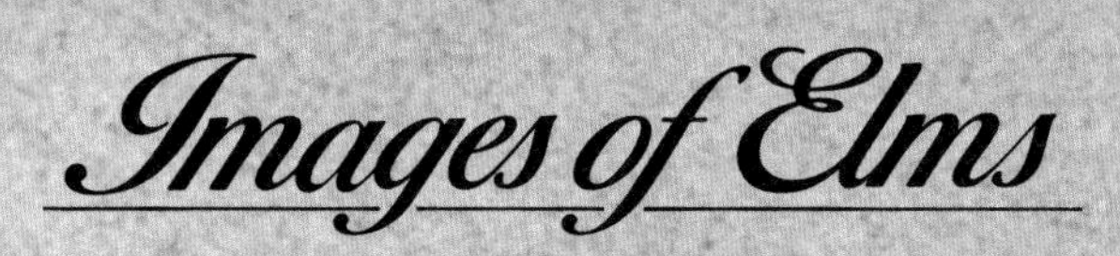

Images of Elms

Let us look at previous generations of elms in verse. They appear in Victorian poetry at moments of stillness – between train journeys, perhaps, which seemed (and often were) incredibly rapid. The elms of this period are again large old trees which appear to be quite permanent. To William Morris, the socialist in retreat at his Oxfordshire manor house, they meant home. But in *Pilgrims of Hope* he adds a curious glint to elm-imagery:

Hark, the wind in the elm-boughs; from London
it bloweth
And telling of gold and of hope and unrest. . . .

Earlier in his life, as a weaver of medieval fantasy, he celebrated Lady Day (25 March) with a picturesque but bloody battle under the elms.

. . . to the crash of the meeting spears
Down rained the buds of the clear spring weather
The elm-tree flowers fell like tears.

Crimson tears for the Annunciation – a new symbolism for elm flowers. It would have to be wych elm in March.

Elms tend to be seen by poets in alternative roles: as witnesses of death, or as large symbols of what is most graceful and enduring in the English landscape. Matthew Arnold places them at focal points: his 'signal elm, that looks on Ilsley Downs' and the youthful Thames; and at *Rugby Chapel*

Coldly, sadly, descends
The autumn evening, the Field
Strewn with the dank yellow drifts
Of withered leaves, and the elms,
Fade into dimness apace,
Silent.

At another still point of the Victorian world, Tennyson immortalized his elms in two lines of daring mastery from *The Princess*:

The moan of doves in immemorial elms
And murmuring of innumerable bees.

It is a country-house picture of elms, and probably was a true one. Yet the colours are too clear: it is only as English as any Pre-Raphaelite pastoral. Browning is more modern, eight years earlier, and his natural history is very good (he describes the elm's epicormic shoots. The elm is at the heart of his feeling for English landscape.

And whoever wakes in England
Sees, some morning, unaware,
That the lowest boughs and the brushwood sheaf
Round the elm-tree bole are in tiny leaf
While the chaffinch sings on the orchard bough
In England – now!

That vignette has the breath of the real countryside, of which John Clare is the true poet. As he wrote, in the 1840s, in Northampton asylum:

Who break the peace of hapless man
   But they who truth and nature wrong?
I'll hear no more of evil's plan,
   But live with nature and her song.

One of *his* elms is *The Shepherd's Tree*; and he is not afraid of its branches falling. Strangely enough, though, the tree makes him think of mortality and forgotten deeds:

Huge elm, with rifted trunk all notched and
      scarred,
   Like to a warrior's destiny! I love
To stretch me often on thy shadowed sward,
   And hear the laugh of summer leaves above;
Or on thy buttressed roots to sit, and lean
   In careless attitude, and there reflect
On times, and deeds, and darings that have
   been . . .

In another, longer, poem, *The Fallen Elm*, he first expresses his love for the tree, the sense of protection it gave, its habit of holding its leaves until the winter; then he connects the felling of this tree with the ruinous effect of the enclosures of common land on the rights of cottagers.

Old elm, that murmured in our chimney top
The sweetest anthem autumn ever made
And into mellow whispering calms would drop
When showers fell on thy many coloured shade
And when dark tempests mimic thunder made –
While darkness came as it would strangle light
With the black tempest of a winter night
That rocked thee like a cradle in thy root –
How did I love to hear the winds upbraid
Thy strength without – while all within was mute.
It seasoned comfort to our hearts' desire,
We felt thy kind protection like a friend
And edged our chairs up closer to the fire,
Enjoying comfort that was never penned.
Old favourite tree, thou'st seen time's changes
   lower,
Though change till now did never injure thee;
For time beheld thee as her sacred dower
And nature claimed thee her domestic tree.
Storms came and shook thee many a weary hour,
Yet steadfast to thy home thy roots have been;
Summers of thirst parched round thy homely
   bower
Till earth grew iron – still thy leaves were green.
The children sought thee in thy summer shade
And made their playhouse rings of stick and stone;
The mavis sang and felt himself alone
While in thy leaves his early nest was made,

And I did feel his happiness mine own,
Nought heeding that our friendship was betrayed,
Friend not inanimate . . .
– Such was thy ruin, music-making elm;
The right of freedom was to injure thine:
As thou wert served, so would they overwhelm
In freedom's name the little that is mine.

Freedom was the hypocritical cry of those who took away the poor man's right to graze his cow and gave him instead a wage only just above starvation level. Land was cleared for profitable agriculture. The connection between the enclosures and what Clare saw as wanton felling is clarified by another poem, *Remembrances*, regretting the destruction of trees which left the fields and streams naked.

By Langley Bush I roam, but the bush hath left
   its hill,
On Cowper Green I stray, tis a desert strange
   and chill,

And the spreading Lea Close oak, ere decay had
  penned its will,
To the axe of the spoiler and self-interest fell
  a prey,
And Crossberry Way and old Round Oak's
  narrow lane
With its hollow trees like pulpits I shall never
  see again,
Enclosure like a Buonaparte let not a thing
  remain,
It levelled every bush and tree and levelled
  every hill
And hung the moles for traitors – though the
  brook is running still
It runs a sicker brook, cold and chill.

Many landlords, of course, replaced the trees and bushes they had cleared by planting hedges, and putting elms in them. But Clare's picture of his native oak forest in Huntingdonshire must be true. A century later some of the lovely disorder had crept back into the countryside, softening the edges of the enclosed land, filling odd corners with bits of wild wood, leaving banks and grassy edges for wild flowers to thrive. This would have pleased Clare, but he would not have recognized any old landmarks such as he mentions in this verse. It is difficult to conceive of a wasteland countryside which was yet as well-charted in local knowledge as the streets and corners of suburbia.

A shorter and later poem to *Autumn* is domestic, not pastoral, and full of sharp observation of his domestic landscape.

I love the fitful gust that shakes
  The casement all the day,
And from the mossy elm tree takes
  The faded leaves away,
Twirling them by the window pane
With thousand others down the lane.

I love to see the shaking twig
  Dance till the shut of eve,
The sparrow on the cottage rig,
  Whose chirp would make believe
That Spring was just now flirting by
In summer's lap with flowers to lie.

I love to see the cottage smoke
  Curl upwards through the trees,
The pigeons nestled round the cote
  On November days like these;
The cock upon the dunghill crowing,
The mill sails on the heath a-going.

The feather from the raven's breast
  Falls on the stubble lea,
The acorns near the old crow's nest
  Drop pattering down the tree;
The grunting pigs, that wait for all,
Scramble and hurry where they fall.

Mossy (it is 'glossy' in another version, but this may be a mistake in transcription) perhaps refers to the green colour of algae-laden bark. But, since Clare is habitually so observant, his elm actually may have been mossy in the cleaner air of the 1830s. His elm in *March* is 'wind-rocked' as the rook prepares her nest, and it is dark green in May, as people get ready to go to Deeping Fair.

The sun was peeping o'er the spreading rows
Of dark green elms alive with busy crows
And round the lodge that darkened neath their
  shade
Loud was the strife that pigs and poultry
  made. . . .

Elsewhere he calls them 'huge homestead elms'. They are important in the village scene in *Pear Tree Lane*:

The elm-trees in our garden wall,
  How beautiful they grew,
Where ring-doves from their nest would call,
  And the vein-leaved ivy grew
At the old house end, with one huge elm
  That turned a whole day's rain;
Storm roared as 'twould the town o'erwhelm;
  'Twas shelter down the lane.

In his madness, far from home, Clare's elms are bathed in remembered sunlight:

The sunny wall where long the old elms grew
The grass that 'een till noon retains the dew
Beneath the walnut shade – I see them still . . . .

And in a later asylum poem they are placed in an ideal setting:

Love, meet me in the green glen
  Beside the tall elm tree
Where the sweet briar smells so sweet agen;
  There come with me,
  Meet me in the green glen.

He forgets his wife, Betty, and believes himself married to his first love, Mary. She it was, no doubt, whom he was to meet under the domestic elm, when he limped from his nursing home, in Epping Forest, to Peterborough, eating on the way one hearty meal – of grass.

An early poem of Wordsworth, *Margaret*, or *The Ruined Cottage*, of some 500 lines, perhaps first established the portentous symbolism of elms. They are introduced at once.

Supine the Wanderer lay,
His eyes as if in drowsiness half shut,
The shadows of the breezy elms above
Dappling his face.

They advertise the proximity of a hidden, overgrown habitation: the cottage, with its well beneath 'two tall hedgerows of thick alder boughs' and 'shrouded with willow-flowers and plumy fern . . . where no one came/But he was welcome . . .' and all 'blessed poor Margaret for her gentle looks'.

After two blighted seasons her 'Wedded Partner' had contracted a fever. He had recovered to a pleasureless poverty – the children could not understand him:

  'Every smile',
Said Margaret to me, here beneath these trees,
'Made my heart bleed'.
  At this the Wanderer paused;
And, looking up to those enormous elms,
He said . . .
'Why should a tear be on an old man's cheek?
Why . . .
  Thus disturb
The calm of nature with our restless thoughts?'

The husband had gone away. The Wanderer, returning once more to 'those lofty elms', heard her tearful tale, roved again, returned again, and waited at the cottage for the rambling widow.

  Deeper shadows fell
From these tall elms. The cottage clock struck
  eight –
I turned and saw her, distant a few steps.

Her face was pale and thin. . . .

Her infant died, and after nine tedious years in unquiet widowhood, Margaret died too.

The poem was begun in 1795, and set in the West Country, not in the Lake District.

George Crabbe, 1754–1832, has been described as the first poet of landscape. He is best known, of course, for his work based on his native Aldeburgh, a place almost treeless on piles of pebbles and sandy causeways. His wife was called Sarah Elmy, but this does not seem to have stimulated any special interest in the trees. In one of his Tales, perhaps autobiographical, he tells of a lover's journey from 'a barren heath beside the coast' to a different, greener countryside, where, at Loddon Hall,

Spread o'er the park he saw the grazing steer,
The full-fed steed, the herds of bounding deer:

On a clear stream the vivid sunbeams play'd
Through noble elms. . . .

which again places our tree in the setting of the landscape garden rather than in the older countryside.

A late poem of Crabbe's is *On a Drawing of the Elm Tree, under which the Duke of Wellington stood Several Times during the Battle of Waterloo*. You hardly need more than the title. It was

A noble tree, that, pierced by many a ball,
Fell not . . .

There are elms at Gray's Stoke Poges, witnesses not of strife, or even of death, only of sleep.

Beneath those rugged elms, that yew-tree's shade
Where heaves the turf in many a mouldering heap
Each in his narrow cell for ever laid,
The rude Forefathers of the hamlet sleep.

James Thomson, whose *Seasons* (1727–30) was much admired by landscape artists even in the following century, is a poet somewhat stylized for our taste, and much addicted to adjectives. His willows were grey, his 'aspin' tall, his elders flow'ring, his hazel 'pendant o'er the plaintive stream'. His oaks were solemn and his elms, predictably, lofty. They were nowhere in the wild, but at

. . . the rural seat
Whose lofty elms, and venerable oaks
Invite the rook, who high amid the boughs
In early spring his airy city builds,
And ceaseless caws amusive. . . .

(He was better at adverbs.) His tree flora is at least wider than that of his contemporary painters, and he is suitably caustic about 'weeping fancy pines'. But, at about 5700 lines, *The Seasons* is too long-winded.

Spencer's and Shakespeare's elms are merely accessories of the vineyard, familiar to them in classical texts which describe how moderately large elms were transplanted to vineyards and there carefully pruned to support the vine without shading it; not the last time two plants have been cultivated together for, at least partly, superstitious reasons.

Spencer's is 'the vine-prop elm' – no more. Shakespeare, in *A Comedy of Errors*, turns it to a nice picture of masculinity.

Come, I will fasten on this sleeve of thine:
Thou art an elm, my husband, I, a vine.

The Elms in Landscape Painting

John Constable (1776–1837) is the only painter ever to have taken elms at all seriously – not only because he was the greatest painter of the English landscape, but because his particular sort of truth to nature depended on an intimate familiarity with his subject. It was for a comparatively brief period, the end of the eighteenth century and the first half of the nineteenth, that painters had ceased to see landscape as the old masters told them they ought to see it, but had not yet begun to see it as a camera likeness. Turner and Constable worked at the height of this unique phase. Though they were almost exact contemporaries, Turner was a prodigy, while Constable was rather slow to develop.

Turner learnt his craft as a topographical artist, travelling about England with his sketchbooks. He certainly knew his trees, but not intimately. There are pages of sketches to show that he had become a Royal Academician (1802) before he had learnt to recognize laurel, horse-chestnut and larch – some typical plantation trees of the gentlemens' country seats he often had to draw. He noted, once, 'polled Arbeles' (abele = *Populus alba* or white poplar) on the Thames, and 'Sycamore . . . Oaks . . . 2 Firs' in a Sussex scene. He drew accurate and beautiful oaks (at Tabley Hall, Cheshire, 1808; and the Cowthorpe Oak, 1815–16), larches (Farnley, 1812), pines (Devon, 1813; Battle Abbey, 1815–16) and beeches (Raby Castle, 1817). There are elms in the background here and there. Turner first visited Italy in 1819. His truth to nature, as is well known, consisted largely of his understanding of the interplay of light, colour and form: this did not lead him to particularize in his painting. However, he left us thousand of pencil sketches of Britain, and one great monument – *Picturesque Views in England and Wales*, engraved in ninety-six plates. These give a true and loving portrait of our beauty spots, towns and people around 1830. But indications of vegetation are in the broadest terms.

Constable's view of truth to nature owed nothing to picturesque preconceptions. 'These scenes made me a painter', he said of his native Dedham Vale: 'The sound of water escaping from mill dams, etc, willows, old rotten planks, slimy posts, and brickwork, I love such things. . . . Painting is with me but another word for feeling, and I associate "my careless boyhood" with all that lies on the banks of the Stour. . . . I had often thought of pictures of them before I ever touched a pencil.'

His landscapes have been described as 'elmy' (by Geoffrey Grigson) and I think this is true. Elms are so important an element of the scenery of Dedham Vale – there are hedges and pollards as well as great trees planted for building barges and lock gates – that their character entered his brushwork. A sketch for *The Lock*, 1826, shows some massive willows on the towpath: these are turned into elms in the painting. The quality of 'elmyness' – a damp-looking, luminous, strongly patterned effect – emanates from even his roughest sketches of Suffolk.

Although his inspiration was a sort of nostalgia, Constable's view of nature was so original that almost no one except his friend Archdeacon John Fisher appreciated it. In the year after the *Hay Wain*, arguably the greatest British landscape painting, was exhibited, but still not sold, Constable had to borrow £30 to keep going. About this time he did the fine study in oils of a group of trees in a hedge near Hampstead, of which he wrote to Fisher:

'. . . independent of my *jobs* I have done some studies, carried further than I have yet done any,

Above **'East Bergholt Church with Elm Trees'. A pencil sketch by Constable.** Opposite **Constable: 'Landscape' (The Cornfield) 1826**

A 'Cornfield with Figures' once ascribed to Constable

particularly a natural (but highly Elegant) group of trees – Ashes, Elms and Oaks etc which will be of quite as much service to me as if I had bought the field and Hedge Row, which contains them – and perhaps one time or another will fetch as much for my children. It is rather larger than a Kitcat and upright.'

('Kitcat' was 36 × 28 inches (90 × 70 cm), as used for portraits of members of the Kit Cat Club.) Fisher wrote to Constable from Salisbury, 12 December 1823,

'the great storm played destruction at Gillingham. It blew down two of my great elms, bent another to an angle of forty-five degrees . . . stripped a third of all its branches, leaving only one [tree] standing entire. This I have taken down, and your wood exists only on your millboards. The great elm in the middle of the turf is spared . . .' (Gillingham is near Shaftesbury).

Constable replied:

'Second. How much I regret the grove at the bottom of your garden. This has really vexed me. I had promised myself passing the summer hours in its shade. Third. I am glad the great elm is safe.'

The other great elms in his drawing of Old Hall Park remain without doubt the best representation of elms in art. They are presumably *Ulmus carpinifolia*, smooth-leaved elm.

Painters before Constable rarely seem to have recognized elms or any other tree. Gainsborough's oaks are often clear enough, but only because he was an honest painter. Ruysdael, Hobemma, Paul Sandby and John Crome all portrayed old oak trees and apparently neglected to paint elms. These artists were all concerned to make true records of their countryside, and it is fairly clear that elms were not typical of wild or picturesque country. Painters were not attracted by 'rural' or agricultural scenes.

Eighteenth-century landscapes are full of trees that might be ash, elm, oak or chestnut, and often have characters of all these. Alexander Cozens, not a great painter but a lively theorist and teacher, at least showed that he knew what elms looked like. How many painters included shreded elms such as he engraved? None.

Claude, Poussin, Van Dyke, Wilson and many Dutch masters produced entirely convincing landscapes in their different styles without, it seems, ever asking themselves what sort of trees they were painting.

After Constable, but not perhaps because of him, natural history got into art. The Pre-Raphaelites and Ruskin made careful studies from nature, and would all, except perhaps Rossetti, have been ashamed not to know what they were drawing. Rubens, who did equally accurate and much more powerful studies two centuries before, probably knew the names of his hedgerow plants, but equally, did not care.

But the camera tells us more of the nineteenth-century landscape than the painters. Even the 'realist' nature-painters of France, Millet and Rousseau, Daubigny and Courbet, often treated their trees as amorphous masses of foliage. One lesser known, Constant Troyon, an enthusiastic cow-painter, gives us a distant silhouette of *Ulmus procera* which would fit into the background of a modern colour photograph without the difference showing.

Opposite Constable: 'Elms in Old Hall Park', East Bergholt. Large pencil drawing with some wash

'Landscape and Cattle' by Constant Troyon, painted about the middle of the nineteenth century

Constable: 'Study of Trees at Hampstead'.
Opposite **Constable: 'Landscape – a Barge passing a Lock on the Stour'. Painted, in 1825 as a companion piece to 'The Cornfield'**

John Constable: a small oil sketch of 'Dedham from Langham', dated 1812. Opposite 'Trunk of an elm tree'. An oil study by Constable

Part 2
A Plain Man's Guide to the Botany of the Elm

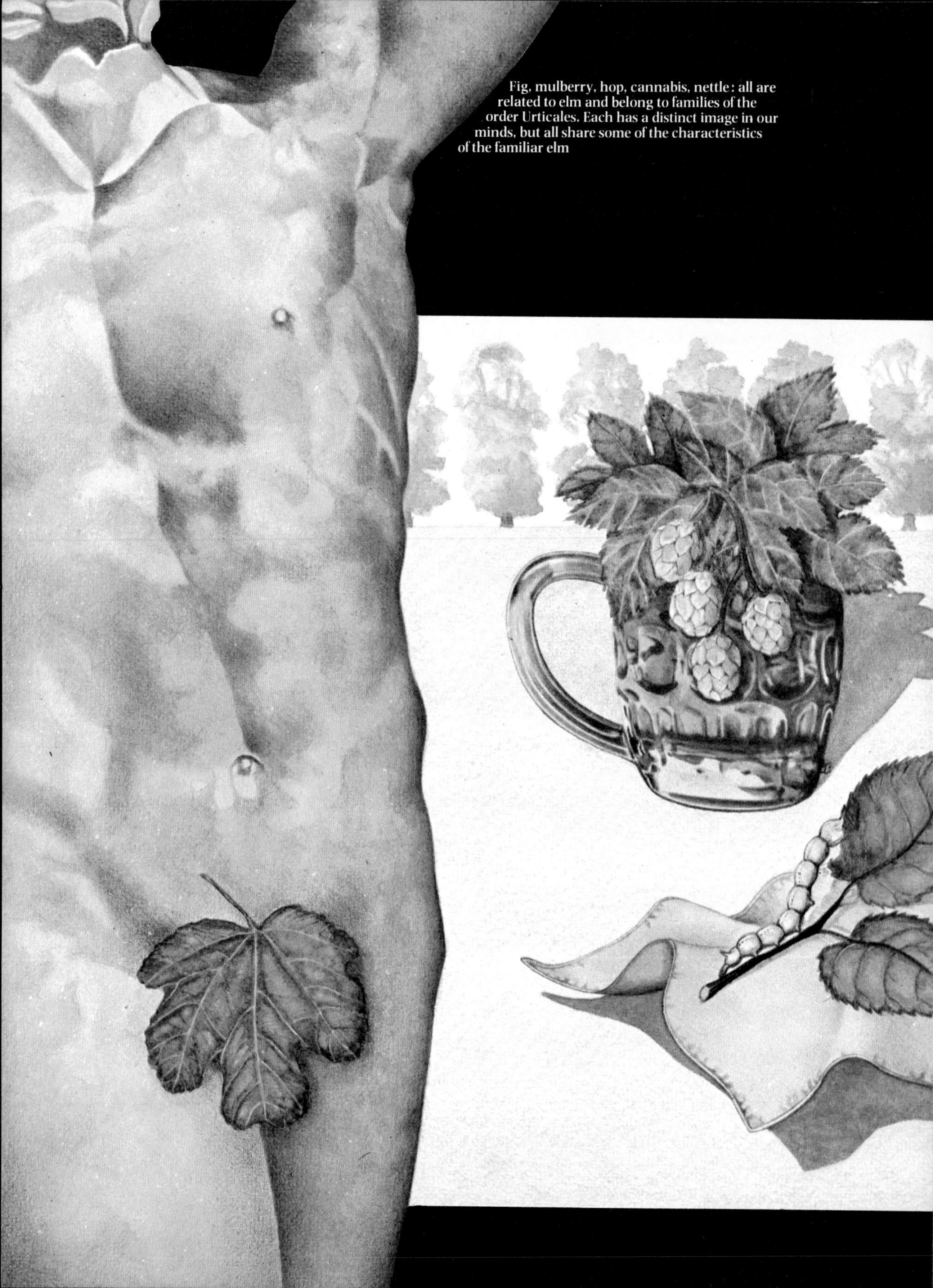

Fig, mulberry, hop, cannabis, nettle: all are related to elm and belong to families of the order Urticales. Each has a distinct image in our minds, but all share some of the characteristics of the familiar elm

Botanically (and zoologically) organisms are classified into orders, families, genera and species. Taxonomy (the naming of living things) corresponds more or less with evolutionary status – usually, I believe, less. It is, as we shall see, a variable and inexact science.

The elms are classified botanically in the order Urticales, containing three families represented in Britain's flora: Urticaceae, Cannabinaceae and Ulmaceae. There are also Moraceae (figs and mulberries), the members of which are mostly tropical and sub-tropical. The Urticales are trees, shrubs and herbs, usually hairy with alternately placed leaves and hermaphrodite or unisexual flowers; the fruits are various.

Plants of this order are highly developed in the evolution of the green world. However, these great flower-bearing plants are comparatively recent additions to the world's flora and they have retained, as shown by the elm and many of its near relations, the 'primitive' vegetative as well as the sexual method of reproduction.

Cousins of the Elms

That 'Elm hateth Man, and waiteth' is purely poetic anthropomorphism, perhaps deriving from the tree's supposed habit of dropping large branches on passers-by, or its frequent use for coffin wood. But a close relation, the stinging nettle *Urtica dioica*, has a sting that is certainly hateful, and some people are said to get nettlerash from the touch of elm leaves. Other relations are the hop, the cannabis plant, and the wonderful tribe of fig trees, only one of which grows in British gardens, along with some mulberries and the graceless osage orange.

The hops' hairy flowers contain lupulin, in a yellow powder, which flavours and preserves beer, but not the Anglo-Saxon ale. The female flowers are more productive, and in this member of the Urticales they are on separate plants from the male, which has led to a minor difficulty in the Common Market of Europe: the continental hops are segregated, ours not. Like the elm, the hop has been called a weed; a wicked weed in the Parliament of Henry VIII. (Heaths, ground ivy, wormwood, yarrow, wood sage and broom were used, with honey, to flavour ales made from fermented malt but not called *bier*). Hops were thought to make one melancholy – it is perhaps not hops but some less wholesome additive which causes some mass-produced beers to make one's legs ache. Hops also caused 'tormenting diseases' and shortened one's life. Thus, they were regarded with great suspicion even by seventeenth-century enthusiasts for 'real ale'. But the pharmacists adopted oil of hop for improving appetites and promoting sleep. Hop tea is made from the leaves of Kentish hops mixed with the now orthodox Ceylon tea to make an infusion both refreshing and sleep-inducing, if that is possible. And warm hops as a pillow are good for toothache and earache – the heat will help if the hops do not.

Elm flowers, which come in February, have no bracts or petals, so they cannot be expected to share the lupulin of the mellow, late-summer hop flower. But they were said to make the bees drowsy. Beer can be made of hops alone, and it can also be made of nettles (with lemon and ginger, the yeast being floated on a slice of toast). This the herbalist calls a botanic beer. Perhaps we might try elm beer, before it is too late?

The tough bine of the hop can be used to make cloth, but it has to be soaked for half a year to separate the fibres. The cloth, once made in Sweden, is coarse and white. Paper has been made of hop fibre, of what quality the paper I do not know. Nettles make both fine cloth and coarse – not only in fairy tales, for captured German overalls in 1916 contained eighty-five per cent nettle fibre, the rest being made up from a tropical member of the nettle family which normally provided gas mantles.

But nettles proved hard to cultivate in spite of being so easy to grow where they are not wanted – they demand a soil rich in nitrogen. Elm fibres are coarser stuff perhaps, for mats and ropes, but they were used by prehistoric Britons and also in this century by eastern Russians and northern Japanese.

Placed with the hop, *Humulus lupulus*, in the family Cannabinaceae is *Cannabis sativa*, long cultivated for the fibres of its stalk. Cannabis is collected from female plants of this tall annual by men in leather coats rushing through, the resin

A great mulberry tree engraved for Loudon, 1838

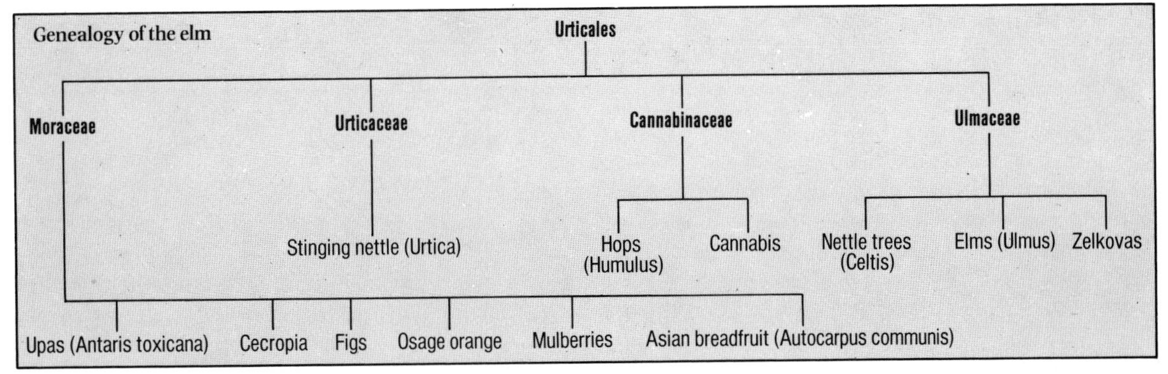

being then scraped off them; in Baluchistan, the shoots are gently rubbed between carpets or rolled underfoot by workers supported by bamboo poles. Hashish, which gave the wild courage and the name to Arabian assassins, is harmless and not addictive – and not very wild-making – but it forms part of the miserable commerce in addictive drugs. Elm trees in dry conditions also exude a resin from the bark; it is not narcotic – though it attracts wasps, bees and flies. The name Indian hemp reminds us that cannabis is primarily a fibre plant: hemp was imported before the last war at the rate of a million tons a year from Russia, for ropes, sacks and sails which are now made of nylon.

Mulberries and Figs

A great family of the Urticales, the Moraceae, contains seven genera mostly of trees with milky sap (latex) and usually dependent on, or symbiotic with, insects. Their relationship with Ulmaceae may seem as obscure as that of the Cannabinaceae, for botany is based on details like the closeness of stamens to ovary or the number of perianth segments. Nevertheless, the similarities and differences are interesting and instructive, and more to the point than comparisons of the elm with neighbouring trees. Elms are never far from nettles in rural England: they are a long way indeed from the banyans and bark-cloth trees mentioned below. But as we shall see, they have been used by men in similar ways.

Members of the Moraceae, 1400 species of them, have 'simple' leaves and unisexual flowers. The Upas tree, *Antiaris toxicaria*, an evergreen, contains a poisonous juice but bears edible fruit. The inner barks of this and an African *Antiaris* provide fibre which is woven into cloth. *Autocarpus communis* is the Asian breadfruit, which was sought by Captain Bligh to feed slaves in the West Indies and is now naturalized there, like the slaves.

Mulberries

The paper mulberries, *Broussonetia*, pithy and juicy, provide paper and tapa cloth from the bark. *B. papyrifera*, still used for beautiful Japanese paper, has large leaves like the (rarely) lobed ones of the mulberry, but with a sandpapery texture.

The mulberries are well known as the food of silkworms, the white mulberry, *Morus alba*, of China, being preferred. Black mulberries grow more easily in England, and are found in the gardens of large houses where often they have been growing since the tree was introduced by the order of James I. The fruit of the black mulberry, the most edible, is red, but those of the white, pink; and red mulberries of the USA are dark purple and fed to pigs.

Mulberries are wind-pollinated and deciduous. The leaves are variously lobed – in the case of the black mulberry, lobes at the base often cover the leaf stalk, just as in the wych elm. The veins underneath are hairy.

Cecropia

Cecropia is a genus of a hundred species of tropical American trees all of which apparently have hollow stems, the homes of ants. The ants protect the tree, and the tree provides food for them in specially developed 'food bodies' at the base of the leaf stalks, so that they will not eat the leaves, which are large, few, umbrella-shaped and downy beneath. The flowers, says Everett (1969), are 'without decorative appeal', male and female on separate trees. *Cecropia peltata*, of the West Indies, is called the Trumpet tree, and is used to make musical instruments. Domestic beasts eat the foliage; *Cecropia* provides a medicinal latex and fibre for ropes. I wonder if those ants would attack *Scolytus* bark-beetle?

Figs

There are hundreds of fig trees in climates warmer than those the domestic fig, *Ficus carica*, can tolerate. They provide fruit so unlike the elm's that discussion here must seem irrelevant, but I cannot resist mentioning a few of this most useful and important family of the Urticales for their vigorous adaptations and endearing habits.

The common fig of English gardens is peculiar in having no male flowers and no need of them. The 'fruit' of fig trees are merely seedless receptacles inside which the petal-less flowers are formed, and the 'fruits' ripen without the necessity of pollination. *Ficus carica* and Smyrna figs, however, only produce edible fruit when pollinated by a special gall wasp. This wasp is often 'cultivated' on a 'caprifig', grown especially for the purpose. Other fig trees each have their own sort of gall wasp, more or less symbiotic.

The shape of the domestic fig leaf is familiar though it is more variable on the tree than in Victorian sculpture. Other figs have heart-shaped or oval leaves and are mostly evergreen. The sycamore of the Bible was the sycamore fig, *Ficus sycamorus*, with its clusters of round fruits, is also called the fig mulberry in North Africa. African figs, like the common fig, grow to astonishing dimensions in central Europe, sometimes developing into circular woods from a single parent – as, for example, does the wonderboom, *F. pretoriae*. Most produce useful fibre, especially the bark-cloth tree, *F. natalensis*, and some sort of latex, typical of the tribe.

The charmingly named *F. exasperata* produces poison and medicine from the leaves, which can also be used as sandpaper. *F. platyphylla*, another largish tree, has leaves a foot long and latex which can be boiled into chewing gum. Many other figs of Africa provide necessities of all kinds.

The banyan tree, sacred to the Hindu, is a fig, *F. benghalensis*. An example of superb adaptation, it is a juvenile epiphyte (growing on another plant without being a parasite), which keeps it out of the way of floods. Its branches spread horizontally and put down roots vertically which thicken into clusters of sinuous supporting trunks. Expanding in this way, says Everett, a single banyan may occupy a large area – up to 2000 feet in circumference and a 100 feet high. The fruits are edible, and the birds drop them on buildings, which are often invaded by roots seeking the ground floor. Perhaps it was the banyan which was known to Gerard as the arched Indian fig of Goa. An evergreen, he wrote, the fruit 'generally eaten, and

Top left **Autumn leaves of the black mulberry**
Top right **Fruits and leaves of the osage orange**
Above **The variously lobed leaves of a common fig tree**
Opposite left **The Florida strangler fig, *Ficus aurea***
Opposite right **The Chinese banyan, *F. retusa***

that without any hurt at all, but rather good and also nourishing'.

The bo tree sometimes appears as a well-shaped rounded tree in the open, 100 feet high. The leaves, with long points, flutter like a poplar's, and are pink when young. The tree is deciduous but has only a very short 'winter'.

The rubber tree is a fig, *F. elastica* of course, the same plant that you can buy in Woolworths in a plastic pot. Its leaf is different from any other fig's. No longer does India rubber come from *F. elastica* in India and Malaya, but from *Hevea braziliensis*, pirated from its native land when export was forbidden, germinated at Kew, and sent in portative greenhouses to Ceylon and Indonesia to start the great rubber plantations.

F. elastica is a strangler fig, starting life in the crook of another tree, and eventually growing into a gigantic evergreen jungle.

The strangler figs form a large group of more or less banyan-like figs. Germinating high in a tree, like mistletoe, they send down vertical roots which encircle the trunk of the host, which usually dies leaving the large crown of the fig tree supported by its basketwork of roots. A common strangler fig in India and West Malasia is *F. caulocarpa*. This has three 'springs' and three 'autumns' every year, separated by a few days of 'winter', while the fruit continues to develop on separate, older branches. *F. variegata* also has flowers and fruit in clusters all over its trunk and branches. *F. geocarpa*, of West Malasia and Borneo, sends out runners as far as 30 feet, from as high as 4 feet up the trunk.

There are strangling figs native to the New World and to Australia. All envelope their host tree in a stockade of snaky stems and slowly kill it by taking all the shade and nourishment.

The only species in the remaining genus, *Maclura*, is the osage orange, *M. pomifera*. One grows in the Cambridge Botanical Garden, and it must have a mate nearby, for it produces fruit. Its relationship with the elms is tenuous perhaps – a thorny native of the Red River Valley of Oklahoma, producing completely solid and inedible oranges. Its leaves are like those of knotweed, only shinier, and it has latex which can cause dermatitis. It is planted widely for it has tough, strong wood, once used for Indian bows and wagon wheels, and said to be the most durable timber in North America.

The Elm Family

The family Ulmaceae comprises the nettle trees, the elms and the zelkovas, about thirteen genera and 150 species of trees mostly native to the northern temperate latitudes. All the trees have alternate leaves which are often asymmetrical at the base. They have clustered, small, unisexual flowers, without petals or sepals. Each floret has a green perianth or cup, which is lobed and contains the same number of stamens as there are lobes – four to eight. The fruits are single-seeded, almost dry berries or nuts, in many species winged.

CHAP. 123. *Of the Lote, or Nettle tree.*

Lotus arbor.
The Nettle tree.

¶ *The Description.*

THe Lote whereof we write is a tree as big as a Peare tree, or bigger and higher: the body and armes are very thicke; the barke whereof is ſmooth, of a gallant green colour tending to blewneſſe: the boughes are long, and ſpread themſelues all about: the leaues be like thoſe of the Nettle, ſharpe pointed, and nicked in the edges like a ſaw, and daſht here and there with ſtripes of a yellowiſh white colour: the berries be round, and hang vpon long ſtalkes like Cherries, of a yellowiſh white colour at the firſt, and afterwards red, but when they be ripe they be ſomewhat blacke.

¶ *The Place.*

This is a rare and ſtrange tree in both the Germanies: it was brought out of Italy, where there is found ſtore thereof, as *Matthiolus* teſtifieth: I haue a ſmall tree thereof in my garden. There is likewiſe a tree thereof in the garden vnder London wall, ſometime belonging to Mr. *Gray*, an Apothecary of London; and another great tree in a garden neere Coleman ſtreet in London, being the garden of the Queenes Apothecarie at the impreſſion hereof, called Mr. *Hugh Morgan*, a curious conſeruer of rare ſimples. The Lote tree doth alſo grow in Africke, but it ſomewhat differeth from the Italian Lote in fruit, as *Pliny* in plaine words doth ſhew in his thirteenth booke, ſeuenteenth chapter. That part of Africke, ſaith he, that lieth towards vs, bringeth forth the famous Lote tree, which they call *Celtis*, and the ſame well knowne in Italy, but altered by the ſoile: it is as big as the Peare tree, although *Nepos Cornelius* reporteth it to be ſhorter: the leaues are full of fine cuts, otherwiſe they be thought to be like thoſe of the Holme tree. There be many differences, but the ſame are made eſpecially by the fruit: the fruit is as big as a Beane, and of the colour of Saffron, but before it is thorow ripe, it changeth his color as doth the Grape. It growes thicke among the boughes, after the manner of the Myrtle, not as in Italy, after the manner of the Cherry; the fruit of it is there ſo ſweet, as it hath alſo giuen a name to that countrie and land, too hoſpitable to ſtrangers, and forgetfull of their owne countrey.

It is reported that they are troubled with no diſeaſes of the belly that eate it. The better is that which hath no kernell, which in the other kinde is ſtony: there is alſo preſſed out of it a wine, like to a ſweet wine; which the ſame *Nepos* denieth to endure aboue ten daies, and the berries ſtamped with *Alica* are reſerued in veſſels for food. Moreouer we haue heard ſay, that armies haue been fed therewith, as they haue paſſed too and fro thorow Africke. The colour of the wood is blacke: they vſe to make flutes and pipes of it: the root ſerueth for kniues hafts, and other ſhort workes: this is there the nature of the tree: thus farre *Pliny*. In the ſame place he ſaith, that this renowmed tree doth grow about Syrtes and Naſamonæ: and in his 5. booke, 7. chapter he ſheweth that there is not far from the leſſer Syrtis, the Iſland Menynx, ſurnamed *Lotophagitis*, of the plenty of Lote trees.

Kkkkkk *Strabo*

Nettle Trees

The largest number of species in the Ulmaceae is contained in the genus *Celtis*, the nettle trees and American hackberries. Most of these are deciduous, but not all. The leaves of nettle trees somewhat resemble those of the elms – they are serrated and rather unequal, but the teeth are absent near the stalk and the structure depends on three main veins rather than the single midrib of the elm leaf. The serrations, though uneven, are simple, not double.

Makins (1936) mentions four species which are hardy at Kew.

Celtis australis Loudon calls this the southern celtis or European nettle tree. It is a native of the Mediterranean lands, but also of south-west Asia, hence perhaps its specific name. It has narrow leaves as long as 6 inches, with long, twisted points and sharp teeth, except at the base, harsh above and hairy below. The hairs are longer on the veins and the leaf stalk, which itself is about half an inch long. The tree is round-topped, grows to 80 feet but can be shrubby. It is reputed to grow for 1000 years. The bark is beech-like, the fruit a berry, brown-red to blackish when ripe. It is supposed to have been the lotus which Homer said was so delicious as to make those who ate it forget their country. The fruit are called 'honey berries' in modern Greece.

The European or southern nettle tree is little touched by insects, and is planted in Italy and France as a street tree. The wood is very elastic, yet will take a high polish. It is, or was, grown in coppices in the south of France, to make hay forks,

The nettle tree, celtis australis

ramrods, whip handles and walking sticks, for those who wanted elastic walking sticks.

C. caucasia This is a smaller tree with downy twigs but smoother leaves, wider, smaller, and less pointed than *australis*; the berry is about the same.

C. occidentalis and *C. laevigata* The former, the Missisippi hackberry, also known as the bastard elm, hacktree, hoop ash, nettle tree, grows as far north as Quebec and Manitoba; the latter, the sugarberry, is native to Louisiana and as far north as Indiana. *Laevigata* means smooth-leaved. Both can be 100 feet tall with dense foliage, and they produce timber known as 'soft elm'. Both are commonly planted as street trees in the south and east of the USA.

C. occidentalis is prone to witches' brooms – twiggy concatenations or lumps caused by unusual stimulation of growth on the branches, the result of attack by insects, a virus, or even a fungus – such as affect the birch and the hornbeam (and once the elm) in this country. The bark is pinkish grey, knobbly and flaky, with corky ridges. At least one side of the leaf is often free of serrations; leaves are dark green above and dull paler green below, the veins silky.

Other less hardy nettle trees are *C. trinervia* of the West Indies and the tala tree, *C. tala*, of Argentina. Both have useful, hard timber. In Africa, Nigeria has a large buttressed nettle tree, *C. adolphi friderici*, and South Africans know the white stinkwood, *C. africana*, with whitish bark, smelly but useful timber, edible fruit and fodder foliage; it is popular as a street tree.

C. bungeana, of China and Korea, has glossy dark green foliage; *C. julianae*, of central China, is large leafed with yellow hairs on the undersides. *C. jessoensis* of Japan and Korea, with a more open crown, has been suggested, says Everett (1969), as a disease-free substitute for the elms. *C. koraiensis* has nearly round leaves and larger fruit than most ($\frac{1}{3}$ inch).

Evergreen nettle trees are exemplified by *C. philippensis* (Taiwan to Australia) and *C. cinnamomea*, the stink celtis, of Ceylon.

Nettle tree paralysis is a South Australian illness: on touching the hairy tree an unbearable pain is felt, followed by numbness affecting the arm or the whole side of the body. But this nettle tree is of a different family – *Laportea*.

Makins adds the water elm, *Planera aquatica*, of the USA. It has a simple ovate leaf, irregularly toothed, rough above and scurfy and downy below, on a very short stalk. The fruit is ribbed irregularly – something between the *Celtis* berry and the winged fruit or samara of the elms. Planera was the old name for the zelkovas, those other near relations of the elm.

Leaves and fruit of the nettle tree, Celtis occidentalis

The Genus Ulmus

Elms have pointed oval, double-toothed, approximately parallel-veined leaves, unequal sided at the base. The leaves have downy patches on the undersides, at the junction of the veins with the midrib-axillary tufts. The leaves of some species are hairy or rough above; others are smooth and shining; all have some hair on the underside.

Although the leaves are correctly described as simple, it is an important feature of the elms that their shoots bear a group of leaves, usually five,

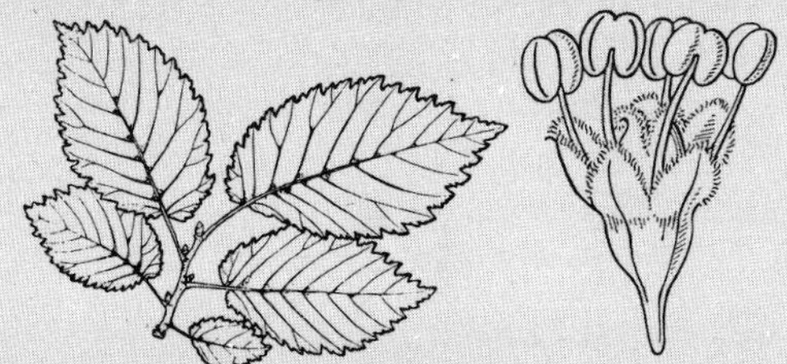

each of which will have a distinct size and character. The terminal bud of most species withers, the largest leaf being formed from the second bud. Buds are fat and pointed, and more or less hairy according to species. Seedlings, rarely seen, have two plain oval seed leaves, followed by opposite, symmetrical leaves.

The flowers appear before the leaves, in clusters. They have no sepals (calyx) or petals (corolla) but consist of a bell-shaped green cup with four or five purplish lobes. This contains the four or five crimson anthers on erect stamens surrounding the ovary with short style and two-pronged stigma.

The elms common in Britain all flower in late February or early March, when a pinkish-brown stipple covers the trees; a subtle flow of warmth into the landscape. Only the hazel catkins precede them, and the elms flower before all the poplars except the native aspen.

The fruit, which forms quickly after the flowers, is flat and only slightly fleshy, nearly circular membrane enclosing the compressed seed, which shows as a darker lump near the centre. The fruit is a samara (winged). The membrane or wing is

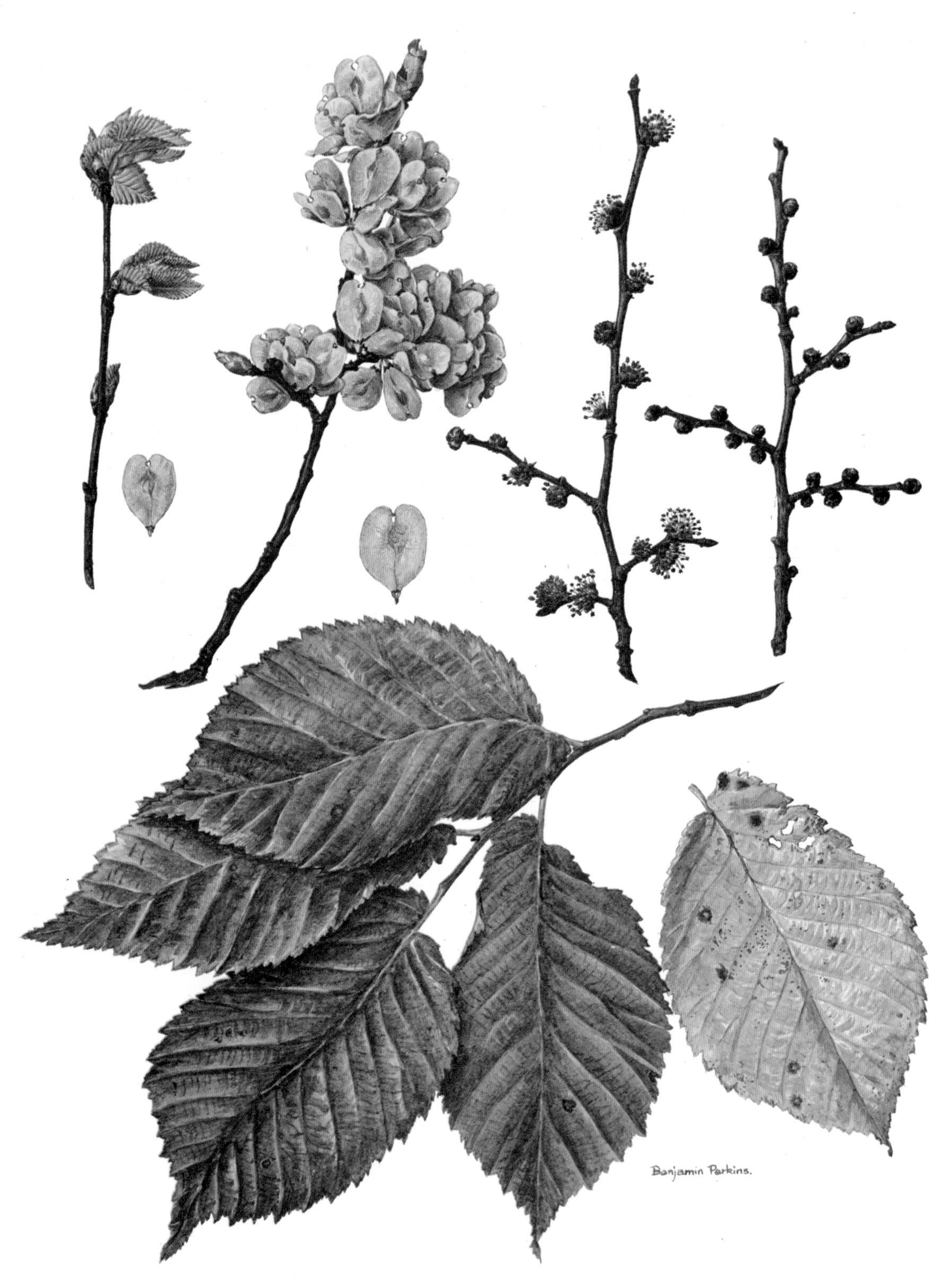

Opening leaves, fruit, flowers, twig and leaves of the Dutch elm, Ulmus × hollandica

Elm flowers, February

notched at the top, with two small hooks – the vestigial stigma – which variously meet or overlap. The clusters of fruit are at first a warm, pale green, giving sometimes the effect from a distance of new leaf. They turn light brown as the leaves appear in late April or May.

Clapham, Tutin and Warburg (1952) say there are about thirty species of elm in north temperate regions and on mountains in tropical Asia. Other authors give the number as fifteen, eighteen or forty-five, no doubt taking different areas into account. The *Flora of the British Isles* must be quoted again before we go any further:

'The British spp. present many difficulties to the taxonomist and are incompletely worked out. Habit of tree and shape of lf appear to provide the most valuable characters for the determination of spp. As in most trees, lvs of sucker shoots or those on rapidly growing branches and on young trees may differ greatly from those on slow-growing laterals on mature trees . . . Lvs from suckers, epicormic or lammas shoots of spp. whose mature lvs are glabrous are hairy.'

The last sentence means that leaves from elms with smooth leaves revert to a common hairy or scabrous form on shoots from the roots or stem, or on late summer growth. Such juvenile leaves are also unusually large and may be exceptionally coarsely toothed. Lammas leaves, appearing about August, replace foliage 'burnt off' in spring gales. Lammas is about 1 August, the seventh Sunday after Trinity.

'Hybrids are of frequent occurence and usually fertile', according to the *Flora*. This is a classic scientific understatement. Hybrids are so common as to make the species comparatively rare, and the commonest of the hybrids are not simple crosses but trees of multiple heredity. This may apply as much to great trees planted 300 years ago as to any elm which springs in the hedge. No wonder the characteristics of 'true' species are hard to pick out. Indeed the species themselves have died out in the central part of their distribution, and only survive in England because they are at the extreme edge of their range. They have little fertile seed, but reproduce their kind by the vegetative process of suckering from the roots. These are the native orphans of parents long since hybridized out of existence in Europe, and they live obscurely alongside the hordes of multiple hybrids often imported and planted by farmers and landowners. These imports again multiply with 'natural' hybrids of the countryside, many of them no doubt the results of ancient selective breeding for specialized purposes.

It used to be thought that none of the elms except the wych elm, native to Ireland, Wales and Scotland, was a true native – the theory was that all the others had been brought by the Romans, to grow vines on, as was their custom. The Romans may have brought their own varieties of this and other trees, but the picture now drawn by the botanists is very different: the English elm, *U. procera*, for instance, wherever it originated, is now confined to the south of England and the north-west of France, where it must have spread in the postglacial period, along with our other native trees. Other species remain only in parts of midland England.

But the true species are more clearly defined in the textbooks than they often appear in the fields and hedges. Similar shapes of tree will often be recognized in any given area because they or their predecessors will have been planted from a single batch of cuttings by a local landowner at the time of the enclosures. They are not likely to conform to textbook examples or even to be a 'recognized' hybrid. Frequently, the dominant character of *U. procera* will be apparent in the rough, small and rather broad leaves, and the domed shape.

Sucker shoots of elm forming an impromptu hedge on Hampstead Heath

The botanical puzzle

All elms with rough, small leaves were included under the name *Ulmus campestris* even in the heyday of botanical 'splitting' of species. Loudon, in 1838, summed them up under *varieties*. He wrote:

'These are very numerous, both in Britain and on the continent; and most of them have been selected by nurserymen for their seed-beds. Anyone . . . who has ever observed a bed of seedling elms, must have noticed that some have large leaves, and some small ones; some are early, and some late; some have smooth bark, and some rough bark; and some soft leaves, and others very rough ones. Some varieties are higher than others; the branches take now a vertical, and again a horizontal direction. In short, while botanists describe, and cultivators sow, they will find that nature sports with their labours, and seems to delight in setting at fault alike the science of the one, and the hopes of the other.'

Some idea of the complexity of studying the English elms with Loudon can be gained from the following examples, all under the heading of *U. campestris* and here slightly abbreviated:

'*U. c. vulgaris* very twiggy; pale smooth bark; with almost horizontal branches. Leaden-coloured bark, splitting into long thin strips with age. A bad variety for timber.
U. c. latifolia broader leaves, earlier.
U. c. alba upright. The old bark cracks in irregular long pieces, and becomes very pale with age. Shoots tinged with red; leaf stalks quite red. Leaves shining and doubly and deeply serrated, bearing a near resemblance to those of *U. effusa* [a separate species, described below]. Valuable timber.
U. c. acutifolia Like the last, but leaves more tapering and branches more pendulous. Common in parts of Essex, Suffolk and Norfolk.
U. c. stricta Red English elm. The growth is very rigid and the tree forms poles of equal diameter throughout. Called *U. c. rubra* in the Horticultural Society's Garden.
U. c. virens The Kidbrook elm. Almost evergreen, with red bark: tree of spreading habit. Like the last mentioned, grows well on chalk. Despite the name Kidbrook, in Sussex, it is a *Cornish variety*. A fine tree in the Horticultural Society Garden is the same, named *U. montana nodosa*.'

Loudon includes the Cornish and the 'freer growing' Jersey elm also as subspecies of *U. campestris*, and ends his list with *U. c. tortuosa*, the French *l'orme tortillard*, the twisted elm.

'The twiggy field, or English, elm' from Loudon

Under 'Ornamental, or curious, trees' he includes the silver-leaved elm which is a clear variety of *U. procera*, known now as '*Argenteo variegata*', two birch-like elms, *U. c. betulifolia* and *U. c. viminalis* (more twiggy and pendulous), both useless for timber, and the Chinese and Siberian elms, A yellow variegated form and two varieties with curled leaves complete this miscellaneous collection, but there are several to be mentioned, known only by young plants: *U. dubia* sounds the least promising of these. Eleven French elms are listed by name and brief description – no scientific names.

Loudon's second 'species' was *U. (c.) suberosa*, the cork-barked elm. He also has *U. (c.) s. vulgaris*, the Dutch cork-barked elm, and five other varieties. Writers after Loudon accepted *U. suberosa* as a species, though Loudon's '(c.)' indicated that he regarded it as a possible variety of *campestris*. It had leaves rough on both sides, more rounded and two or three times as big as those of *U. campestris*. It is now recognized that the Dutch elms, hybrids between *U. glabra* and *U. carpinifolia*, often have corky wings on their twigs, and that corky twigs

LVIII. c. a.

Úlmus (campéstris) màjor.

The greater, *or Dutch cork-barked*, Elm.

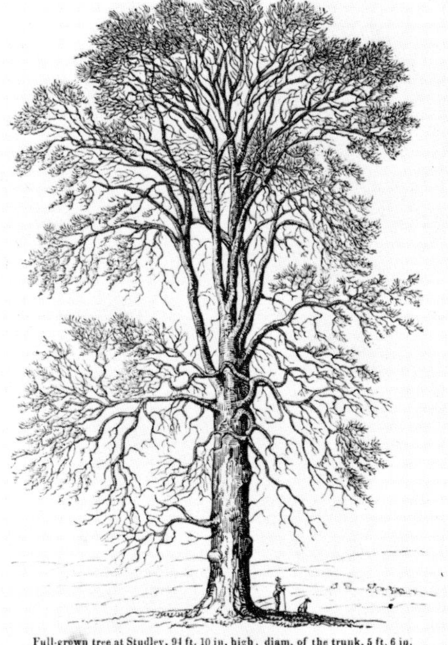

'The greater, or Dutch cork-barked elm' from Loudon

and bark on small hedgerow elms (predominantly *U. procera*) are common. Loudon even has a greater, or Dutch, cork-barked elm, *U. (c.) major* (his third species). This 'appeared to be the kind brought over by William III, from Holland': planted near Kensington Palace and 70 feet high in Loudon's time – perhaps they were the trees of the avenue of Broad Walk, cut down at the end of the Second World War – the trees having survived the Blitz and the V-bombs only to be attacked as a further threat to Londoners' lives and limbs.

Loudon's fourth 'species' was *U. carpinifolia*, the hornbeam-leaved elm, which he knew little about except that it had been found 'four miles from Stratford-on-Avon, on the road to Alcester'. His fifth was *U. effusa*, the spreading-branched elm. It had flowers with long drooping peduncles and hairy fruit, large bright green leaves from sharp, long green buds. Mitchell (1974) accepts this as synonymous with *U. laevis*, the European white elm, which has dark orange-brown buds and is rare in this country. Loudon's *effusa* was propagated here by grafting to *U. glabra*: it was a native of Russia, also found in the forests of Soissons and other parts of France. Its seeds never ripened here and it produced no suckers, otherwise we should have to look for its character in our field hybrids.

Numbers six and seven were *U. montana* and *U. glabra*, the Scottish elm or wych elm and the smooth-leaved or wych elm, with a total of eighteen varieties, some, such as *U. (m.) g.* 2 *vegeta*, the Huntingdon elm, being quite recognizable (*vegeta* is in fact a clone* of the Dutch elm group). Others are interesting, but leave us in some doubt: *U. (m.)* 6 *nigra*, the black Irish elm for instance, with smaller leaves, or *U. (m.) g.* 8 *pendula*, the Downton elm, described as singular and remarkable but with no details except that it was a weeping tree, some sort of cross between the Scotch and the English elms raised from seeds obtained in Nottinghamshire in 1810.

The rest of Loudon's elms were foreign: *U. alba*, a native of Hungary, *U. americana*, *U. fulva*, the slippery elm, and *U. alata*, the wahoo elm. He adds a paragraph on 'Doubtful sorts of *Ulmus*' in which he says, 'we by no means have been able to draw up this article [on the elms] in a manner perfectly satisfactory to ourselves'. I know how he felt. Several years' study are required, as he says.

The situation has now been very much simplified and rationalized by the work of Melville at Kew. We now have these native elms:

1	*U. glabra* Hudson	wych elm
2	*U. procera* Salisbury	common or English elm
3	*U. coritana* Melville	Coritanian elm
4	*U. plotii* Druce *(U. minor)*	Plot's elm, lock elm
5	*U. carpinifolia (U. nitens)*	smooth-leaved or feathered elm
6	*U. angustifolia* Weston	Cornish elm

Hybrids of 1 and 5 give the various members of the Dutch elm group; 3 and 4 give us *diversifolia*. Jersey and Wheatley elms are subspecies of *U. angustifolia*, of which the Goodyer elm is also a subspecies.

That is the situation on paper. In the field it more resembles the picture drawn by Loudon.

*Clone – all the individual plants produced vegetatively, either by horticultural cuttings or by natural suckering, from a descendant of a single pair of trees. Thus, all Huntingdon elms, for instance, have the same two parents, and a specimen of, say, *U. procera* in the wild may have the same two parents as a prehistoric ancestor, itself possibly a hybrid.

Hybrids are sexually produced and in trees are fertile, often more fertile or adapted than either parent.

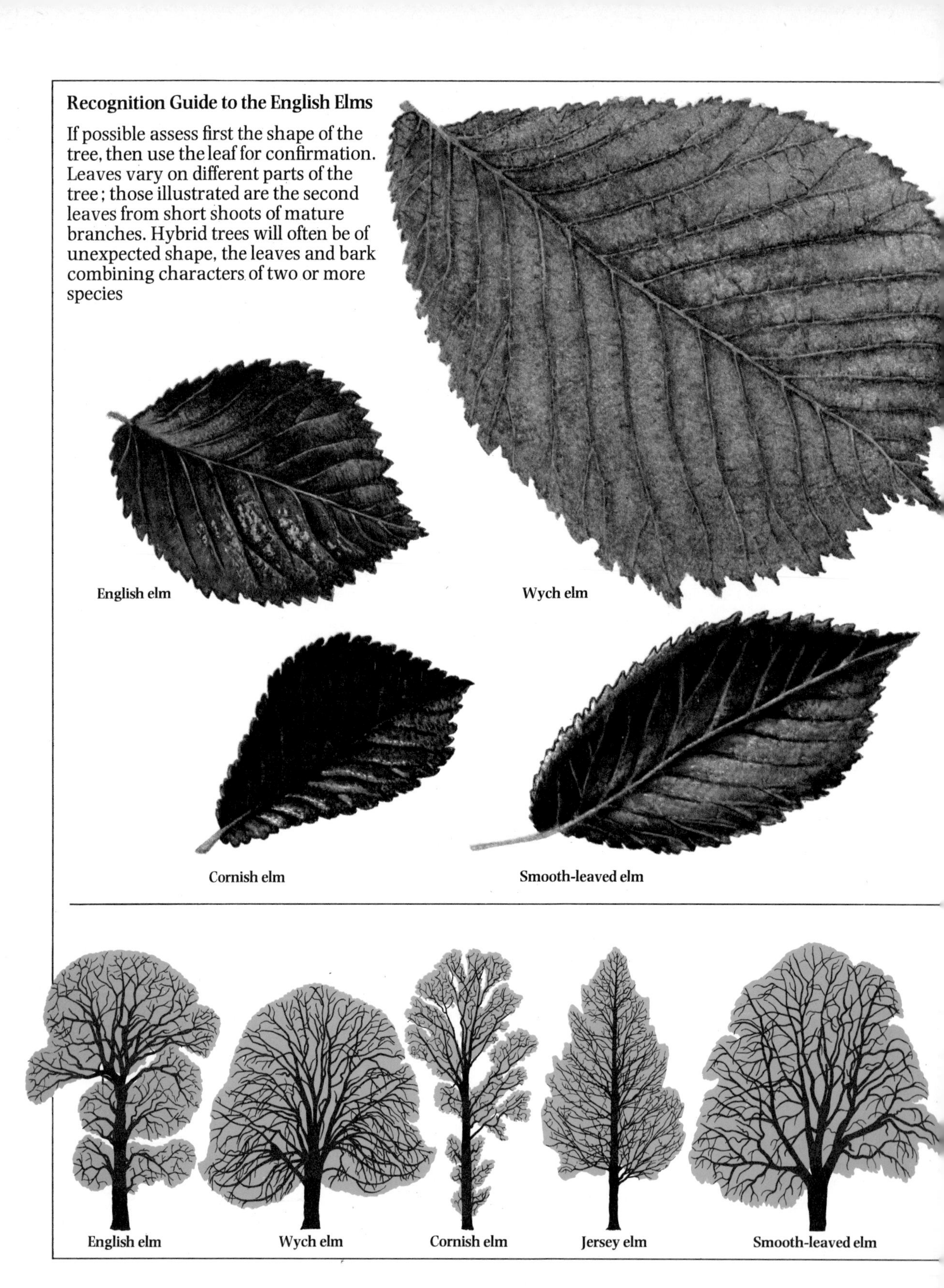
Recognition Guide to the English Elms
If possible assess first the shape of the tree, then use the leaf for confirmation. Leaves vary on different parts of the tree; those illustrated are the second leaves from short shoots of mature branches. Hybrid trees will often be of unexpected shape, the leaves and bark combining characters of two or more species
English elm
Wych elm
Cornish elm
Smooth-leaved elm
English elm
Wych elm
Cornish elm
Jersey elm
Smooth-leaved elm

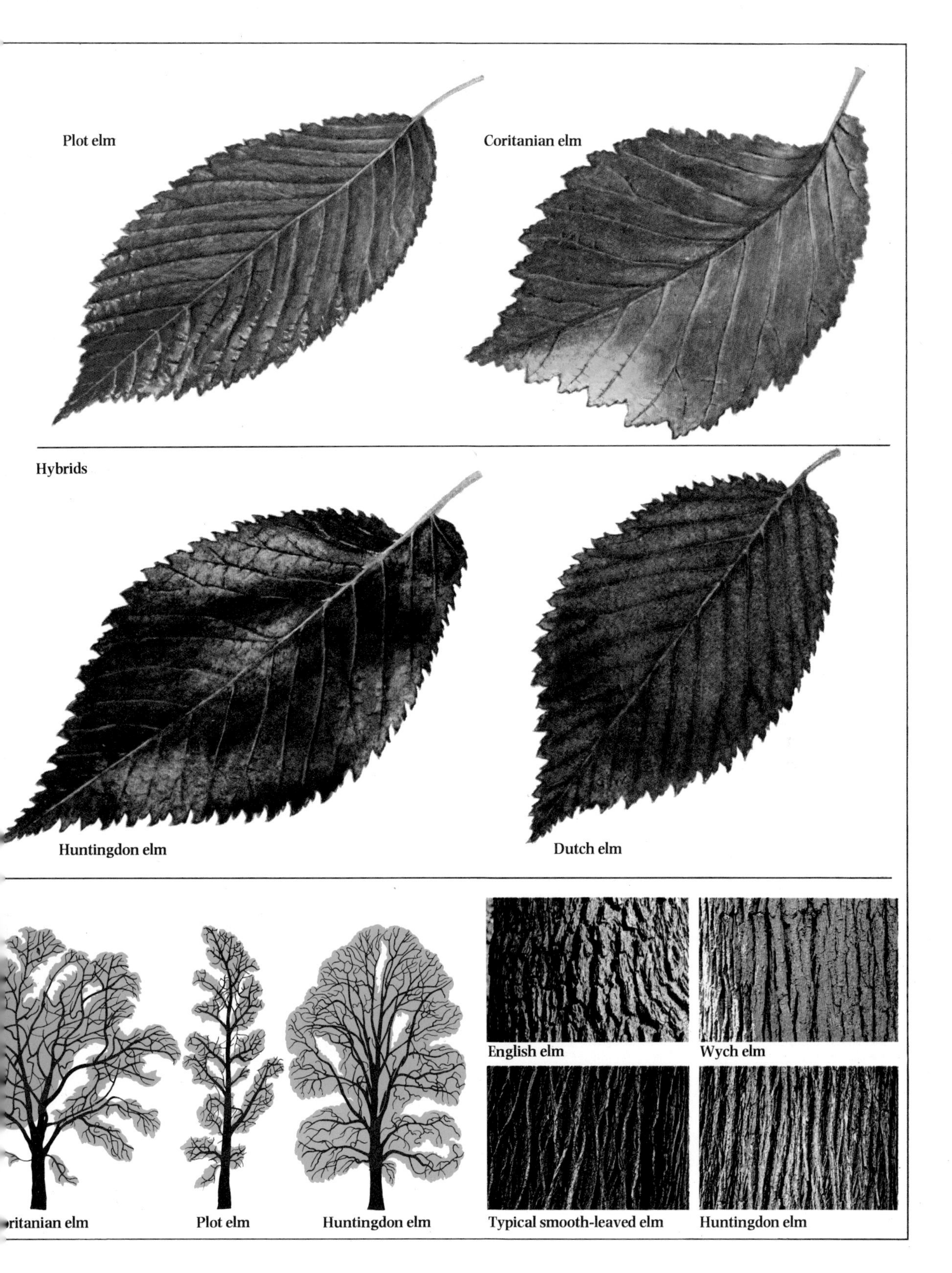
Plot elm
Coritanian elm
Hybrids
Huntingdon elm
Dutch elm
ritanian elm
Plot elm
Huntingdon elm
English elm
Wych elm
Typical smooth-leaved elm
Huntingdon elm

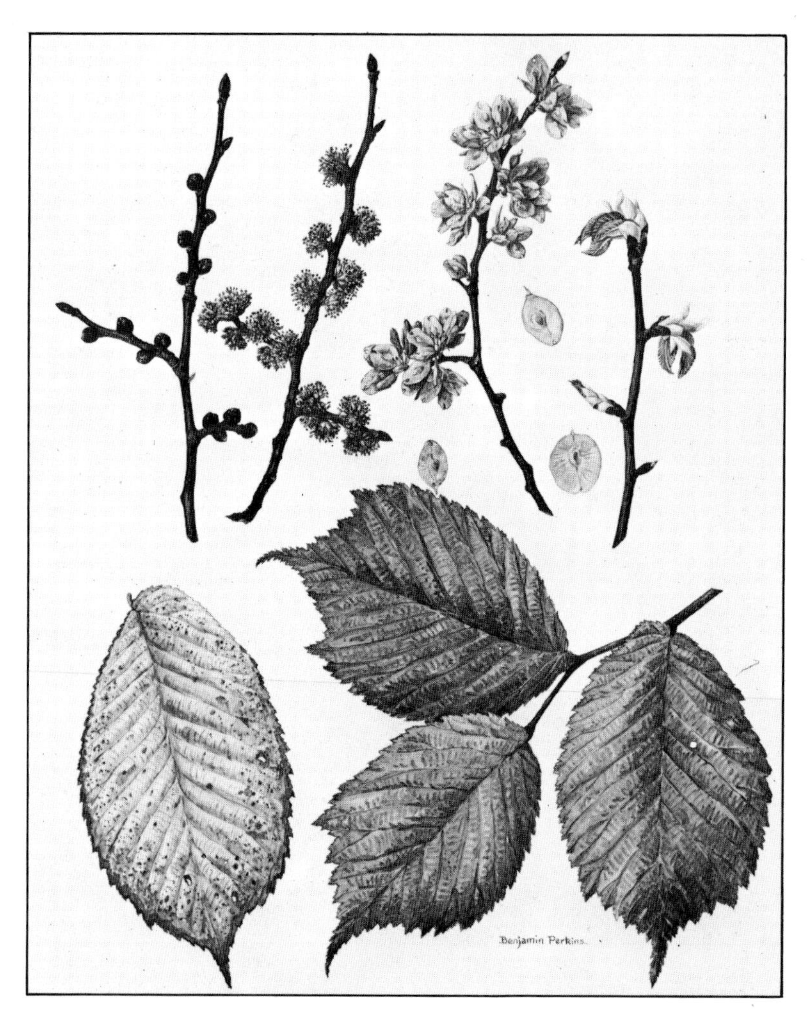

The Wych Elm

Ulmus glabra grows into an irregular rounded shape and can be 125 feet high. The branches begin to fan out and upwards at a comparatively short distance above the ground. Some lower branches sweep low almost to the ground then turn upwards, while the upper ones are also sinuous. Multiple stems are not uncommon, but these may be ancient coppice trees or have grown from two adjacent seeds. The bark is smooth at first on young branches; this justifies the name *glabra*. The colour of the young bark varies according to locality. It can look black in damp northern valleys, or dull silver grey, or a shining brownish-grey near the sea, where it may be flecked with light grey patches. Light and dark wavy lines form towards the trunk, where they become fissures and ridges on the old bark.

The bark of old trunks is brownish-grey, always paler than other elms. The ridges are always recognizable: parallel, usually twisting, occasionally but not frequently broken, and sometimes stringily overlaid. They never become the deep, massive furrows of the oldest smooth-leaved or common elms. In many of its northern and western habitats, the bark of *U. glabra* is touched with the white of incrusting lichens – as are other trees of course, but rarely the southern elms, which only go green with algae. Even the Cornish elm in its presumed native environment rarely carries lichens on the trunk. There are sometimes a few suckers from the trunk of the wych elm, especially in open situations.

The branches divide at narrow angles until, characteristically, they produce shoots almost at

Opposite **Twig, flowers, fruits and leaves of the wych elm**

Majestic wych elm at Chilton Foliat, Berkshire in early summer

right angles: only the much finer twigs of some hybrid Plot elms are similarly placed. Wych elm twigs are comparatively stout and hairy, but after two to five year's growth they become smooth. From a distance, their stoutness will not be apparent against the sky. The shoots are more or less pendulous on upper branches, but spreading near the ground.

The flowers are large for an elm, with crimson to purple anthers on white filaments (purplish in other elms) which gives a pinkish effect. They are followed in three weeks by golden green fruit, $\frac{3}{4}$ inch long, often very abundant. The notches of the fruit often end in tiny crossed horns, but these are variable. The seed is usually fertile, distributed by wind. The seed leaves (cotyledons) are broad-elliptic, with ears at the stalk.

The leaf is pale bright green, large – from $3\frac{1}{2}$ to 6 inches – rough above, downy to rough below, on a very short leaf stalk (petiole) which is covered by the round ear of the long side of the leaf. If the leaf stalk is longer than $\frac{1}{4}$ inch it is not a wych elm but probably a Dutch elm, the other large-leafed type.

U. glabra has always been clearly recognized, but the development of scientific botany led to a certain amount of confusion. In 1768, Miller coined the name *U. scabra*, his *U. glabra* being our *carpinifolia*. Loudon (1838) established the *U. montana*, the mountain, Scottish, or wych elm. To older authors it was the wych hazel, used for bows by the ancient Celts, and as a protection against witchery by Midland milkmaids, who had a notch cut in their churns where they fixed a sprig.

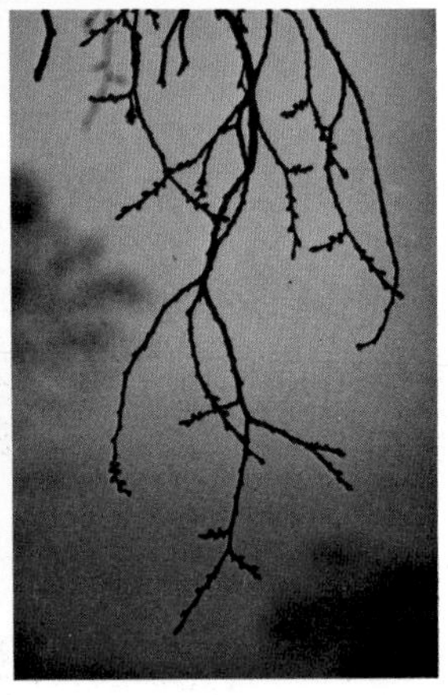

Ulmus montàna fastigiàta.

The fastigiate mountain, *or Scotch*, Elm.

Above left **Typical young leaves of wych elm, a mature divided trunk, and pendulous twigs in winter.** Above **Loudon's 'Ulmus montana'.** Below **shouldered leaves from Dumfries (2) and Cornwall.** Opposite **The wych elm as the hedge tree of the north**

Whatever is the origin of 'wych' it was nothing magic – the confusion was etymological . . . and yet superstitions have often a very long history. (Commercial witch hazel, sold for treating bruises, is a decoction of the bark of *Hamamelis virginiana*: spotted alder, winterbloom or snapping hazelnut.) Loudon takes the wych to mean spring, quoting Droitwich and other spas: but 'wick' and 'wych' in place names usually mean a farm or a dairy farm, which brings us back to those milkmaids. Could it have been the 'farm elm', used for plough and tool handles?

Variations

Some authorities have distinguished between a northern and western type of wych elm and a southern one.

The variation in leaf – and it is not much more than that – is now listed under two subspecies: *Ulmus glabra* ssp. *glabra*, with twigs hairy for up to five years and broad, sharply toothed leaves; *U. glabra* ssp. *montana*, with twigs smooth after two years and narrower, less sharply toothed leaves.

Mitchell (1974) notes that some wych elms have a 'shouldered' leaf shape, which he says is a continental form. The effect is sometimes almost of three-lobes. The leaf reproduced here comes from a burnside in the mountains of Dumfries, and I have found similar forms in Gower, in the hedges of West Kerry and in north Cornwall;

but I will not pretend that any of these trees could not have originated from garden specimens. No other elm leaf shows this three-pointed shape – it would have been a good excuse for the old name wych hazel had it been the native form, because the leaf of the hedgerow hazel often takes this jagged shape.

The distribution of *U. glabra* in Britain, if we ignore subspecies, is wide: thin in Dorset like all elms, and in central Ireland and the south-west peninsula of England. In north and central Wales and north of the Mersey in England it is the prevalent elm species, and it forms thick hedges in Suffolk. In highland Scotland it is the only elm. Here it is a tree of the waterside, growing alongside alder: it spreads along streams, and is not found at any great height unless very well sheltered: it is always well below the line of birch. It forms small woods in the glens, with rowan and bird-cherry. Isolated old trees left at the edges of spruce and larch plantations suggest that considerable numbers have been removed, and Edlin's (1956) figure of 32 000 acres of elm woods in Britain, mostly made up of wych elm, is probably no longer true. Scottish wych elms are conspicuous in autumn by their brilliant gold colour: not the result of any special quality, but a result of cleaner air than most of us are used to. Another characteristic is the abundant lichen and moss on Scottish trees.

Edlin gives us the Celtic and Gaelic names of elm, *llwyfan* and *leamhan*, suggesting the latter appears as *leven* in Loch Leven. Wych elm is the only elm native to Ireland: it grows with fuschia in the hedges of Kerry and with hazel in the secret valleys of County Clare. It has some preference for limestone; there are fine specimens by the river at Kendal, and in the pastures of the Lune valley. It can be found in mixed woodland, and even in the scrub of the Chiltern escarpments. It is confused, and pollarded, with various Dutch elms in Suffolk, and in old coppices is in a minority among hazels or hornbeams, many-stemmed on a great mound or stool, and spreading its branches lugubriously among its neater neighbours.

Wych elm wood will split, while logs of other elms resist the axe: its uses and more of its history I will relate in Part 4.

Cultivars

Distinct from subspecies are some cultivated varieties, familiar and rare. The best known is *pendula*. It dates from 1816 and is the traditional weeping wych elm of churchyards and grand gardens. This can be a large tree with a solid, heavily burred bole, the branches arching, shoots sweeping the ground but rarely hanging vertically. The branches show like bones on the top of the tree, where they form perches for pigeons, who do not like foliage. The vigorous shoots form arbours beloved of Victorian gardeners, and the shape of the tree, to quote Loudon, 'is admirably adapted for particular situations in artificial scenery'.

A weeping form grafted to Dutch elm stocks 'weeps' only as low as the graft, producing a most architectural mushroom, elegantly massing to one side or the other, even in situations sheltered from strong wind.

The more compact pendulous wych elm to be seen in parks and gardens is the Camperdown elm, with a tortuous, domed head grafted on a straight bole. It is a spreading tree, but not sprawling like *pendula*, and it will not grow tall. The branches are sometimes very twisted, like a bunch of snakes, but always with a fine sculptural quality and a knotty logic of growth. The leaves are larger than normal, and of a pale green colour. The name *camperdownii* comes from Camperdown House near Dundee, where it was discovered in 1850.

The Exeter elm or Ford's elm, *exoniensis* or *fastigiata*, approaches a somewhat pendulous silhouette when fully grown, but is upright as a young tree, with inter-woven, apparently knotted, bunches of leaves on very twiggy branches. The tree forms a goblet shape. Grafted on a tall stock it is hardly attractive, but very compact (for streets); and older trees are undeniably fascinating. The rough, folded leaves remain deep green until late in the autumn; the whitish veins are reminiscent of the coarse sucker shoots of some hybrid elms.

There is also a variety, *crispa*, with large, curling leaves; it is a tall and slender tree with few branches.

Opposite **The weeping wych elm variety pendula has bony-looking branches above the foliage.** Top left **Pendula is rarely symmetrical.** Top right **The smallest variety, nana.** Centre left **A golden-leaved variety, lutescens.** Centre right **Vigorous fruiting of a weeping wych elm in a London park.** Above **Camperdownii is compact but tortured within (see previous page).** Right **Twisted leaves of nana**

The golden-leaved variety of *U. glabra* is *lutescens*, or as they have it at Kew, *macrophylla aurea*. The leaves are bright yellow at first, somewhat heart-shaped, varied in size and appear clustered. The tree is well-rounded and can no doubt develop into a typically shaped wych elm. Mitchell (1974) says it has been planted on ring roads and housing estates. I wonder how it has fared in the present epidemic of disease?

There is a purple-leaved variety which I have not seen, *purpurea*.

U. glabra cv. *nana* – borrowing the specific name, but not the habit, from the dwarf birch – is a small tree with small, curling leaves clustered on all the branches, which are again somewhat tortuous. The specimen at Kew has a neat igloo shape; it is like an attractive scale model of a wych elm, until you enter the low crown to examine the branches, which look like a Japanese brush drawing of a cherry bough.

The Common or English Elm

Ulmus procera means the tall elm: the 'outline billowing like a thundercloud', says Mitchell. This lowland field elm made rural England unique, for it is found nowhere else except Brittany and possibly south-east France. Perhaps it can be a little thinner than a cloud, but its crown is rounded and the branches, approximately even in size, form a closed, not an open, canopy. The lower branches from the typically straight, thick trunk are relatively few. There is a tendency for a mass of foliage to form halfway up the trunk, giving the familiar skirted effect – especially where the lowest foliage is cropped to a level ceiling by cattle. The proportions of a well-grown tree are stately: one-third domed crown, two-thirds trunk with foliage. Occasionally trees or groups of trees have been regularly cropped up the stem, leaving them comparatively naked with round, heavy heads. This habit of shreding* was once very common. The foliage was fed to cattle and the shade cast by the tree was reduced.

The partiality of grazing animals for elm leaves may have very much affected its history: from the Elm Decline in 3000 BC to the enclosures of the eighteenth century, it held its place with difficulty until the hawthorn hedges provided a narrow, but protected, habitat. It will invade the edges of woods but it is not a typical woodland tree.

Not every field elm you see is the pure *procera*, but it is the main character in many complex hybrids which have been planted.

* The proper spelling has two *d*s, but old authors often used only one, and it may suggest a different process, with a possible verbal link with shrouded (to shroud is to lop, says my dictionary).

A young English elm in November

A rounded form of Ulmus procera common in the area of Wychwood Forest

The leaves are rough, rarely 4 inches (10 cm) long on terminal shoots, and less than 2 inches (5 cm) on side shoots, varying, in the pattern of elms, according to their place on the shoot. They are oval to roundish, with short points, and vigorously double toothed, the teeth arching forward. The leaf stalk is less than $\frac{1}{4}$ inch (6 mm), hairy on twigs, which are also densely and persistently hairy.

Suckers from the stem (epicormic shoots) are numerous and at an angle of 60°. Root suckers form a hedge, and appear at unexpected distances from the parent. Sucker leaves show the coarse juvenile form, which persists in young trees for several years. Twigs may be corky.

Young trees are conical in outline, with the straight trunk of their species, the branches rather regular. The bark of old trees is brown-grey, varying to black in towns, and thick, furrowed and broken in a lumpy squarish pattern.

The flowers are dark red, the fruits small ($\frac{1}{4}$ inch), scarce and rarely if ever fertile in the field – if they are, the seedlings are soon eaten. The general opinion is that they never do germinate and the tree reproduces entirely by root suckers, thus avoiding natural hybridization. Edlin (1956) states that the seeds are viable for only a week or two, and that the seedlings must have moist bare earth for their first growth. Experiments have shown that fertility falls sharply three or four days after full ripening, while storage, dry, at temperatures little above freezing point, is necessary to preserve fertility until the next year.

U. procera is native to Britain: it is the orphan of continental parents fatally fertile and hybridized out of recognition. It likes good soil and warmth, and is uncommon north of Yorkshire and Lancashire. West of Plymouth it is rare, but planted near to houses. It is not common in East Anglia, being replaced, theoretically, by *U. carpinifolia*, in fact by various hybrids.

Trees can be over 100 feet. One of 150 feet is on record, with a girth of 20 feet. This would be about 300 years old.

Argenteo variegata is the oldest established decorative variety, the leaves bordered and splashed with white. The form of the tree sticks to type, and great specimens can be seen at Kew, Kenwood (near the west gate) and in many large gardens. Its suckers are also variegated. *Vanhouttei* is a golden-leaved form. *Viminalis* is a narrow tree with very deeply toothed leaves, narrow at the base – perhaps not *U. procera*.

Opposite **The landscape we have lost.** Above **Leaves, fruit, flowers, twig and leaves of the English elm, Ulmus procera**

Corky bark of Ulmus procera.

The silver-bordered variety argenteo-variegata

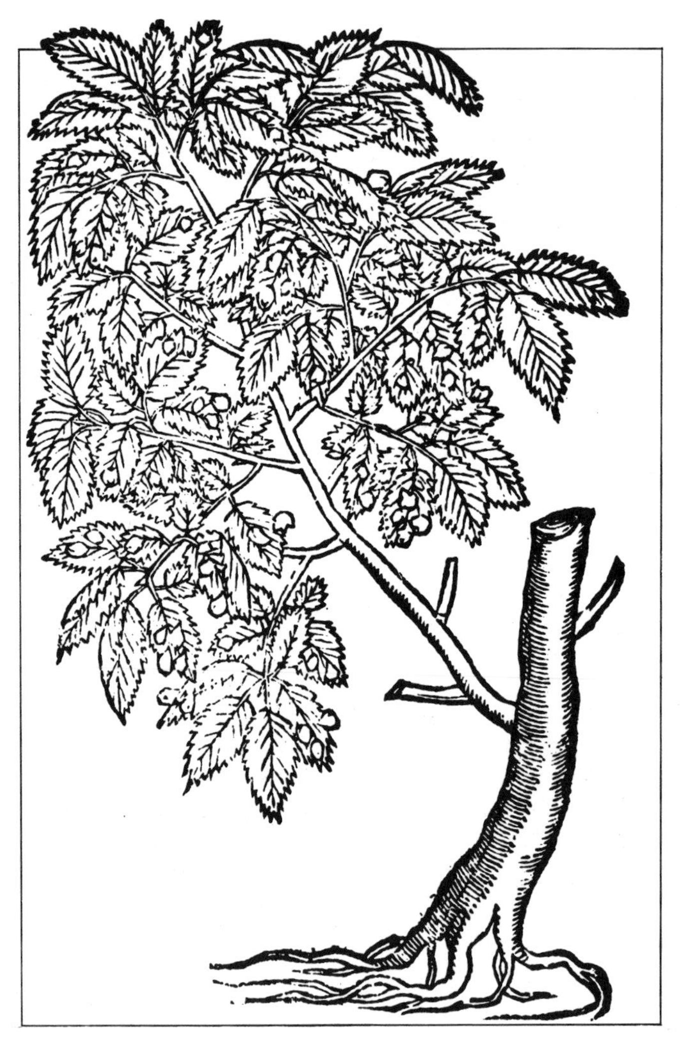

Native Smooth-leaved Elms

That we can pride ourselves on four native species of smooth-leaved elm is because of the exact and exhaustive work of Dr Ronald Melville whose accounts and records have been heavily drawn upon here.

The Cornish elm

The first of the smooth-leaved elms to be identified in England, typically of elm taxonomy, was called 'rough narrow-leaved' – *Ulmus minor folio angusto scabro*. It was discovered by John Goodyer, a botanist of Hampshire, in 1624. Goodyer assisted Johnson, the editor of Gerard's *Herbal* for its second edition in 1633. He wrote that he had seen this elm, 'growing but once, and that in the hedges by the highway as I rode between Christ Church and Limmington'. Cuttings from this and 'the first kinde' (the English elm) he grew side by side – they were 'easy to be discerned apart by any that will look on both'.

These plants were 10 years old and 10 feet (3 m) high in 1633 when he wrote his piece. It was not understood at that time that smooth-leaved elms have rough-leaved suckers, and so his description was wrong on this important point. If he had noticed, when he took his cuttings in 1624, that the parent trees had smooth leaves, he had forgotten it.

From his description, the name *Ulmus minor* was taken by later botanists, notably by Miller in 1731, and applied to a number of different hybrids. The small-leaved elm became a textbook species for two whole centuries. It no longer grew in coastal Hampshire, but in various parts of the

Ascending branches with numerous epicormic shoots, and a rounded crown, are typical of Cornish elms.
Opposite **Gerard's 'Ulmus minor folio angusto scabro'**

Midlands and East Anglia. Usually it was identified by specimens of starved sucker leaves from almost any elm. Sometimes it was the 'lock elm' (which locks the saw). Beginners were warned that its leaves were not all small. Eventually it became, in 1939, *U. diversifolia* Melville: a distinct species with all the prestige of modern science – but only for a few years. It has now retreated into the limbo of hybridization, and they call it *U. coritana* × *plotii* at Kew. This may sound sad – but it is a pretty tree, and a healthy one at the time of writing.

The Cornish elm was at some time a candidate for the title of *U. minor*, and Goodyer's elm somehow survived in the pages of obscure botanical works. Did it still survive in Hampshire? In one of those golden summers of the late 1930s Ronald Melville set off to find it. He found no unusual elms at Goodyer's old home at Maple Durham, near Petersfield – though there were a derelict house and an old, walled orchard, and some English elms. He started from Lymington on the road to Christchurch. A mile out of Lymington a number of elms of an unfamiliar kind met his eye. The foliage of younger trees agreed with Goodyer's 'rough narrow leaves'.

The parent trees were in fact Cornish elms – then *U. stricta* – but with untypical broad crowns and leaves also broader, especially at the base. Thus the Cornish elm was proved to have been known to botany since 1633.

The title of *U. stricta* Lindley dates from the early nineteenth century. *U. cornubensis* also was known, as a horticultural variety, to Loudon, as was var. *sarniensis*, the Jersey elm. The Cornish

Wheatley elm in flower at Golders Green, NW London

The Jersey Elm at Kew Gardens

elm appeared in the 1952 edition of Clapham, Tutin and Warburg, thus:

U. stricta Lindl. var. *stricta* (var. *cornubensis* (Weston) Airy Shaw)
var. *sarniensis* (Loud.) Moss
var. *goodyeri* Melville

Its name was changed to *U. angustifolia*, since var. *goodyeri* was by no means *stricta* in habit. *Angustifolia* means narrow-leaved: it is not really a very good name, though even varieties with broader leaves do have their leaves folded more or less in a boat shape, which makes them look narrow. Considering its largely coastal distribution *U. navis*, or *U. maritima* would have been pleasant.

Angustifolia is really the most recognizably different of all the elms. There are two important forms: the usually coastal and peninsular Cornish elm (var. *stricta*) and (in England) the park and street-side Jersey elm (var. *sarniensis*). The photographs will establish their shapes without long descriptions. It should be noted, however, that in its normal habitat the Cornish elm often has a flat, wind-cut top. Also, *all elms* – and oaks – exposed to great draughts of sea air and unlimited light put out shoots along their trunks and branches, giving them the thick, furry outline which is characteristic of Cornish elms. Indeed all woodland trees when exposed by clear felling to unwonted air and light acquire this clothing of epicormic shoots – not so thickly, though, as does the Cornish elm. As Melville has pointed out, it is characteristic of the species and occurs even when the tree is in a protected situation. The fact remains that other species of elms planted in Cornwall often look decidedly Cornish in outline.

The leaves of the Cornish, Jersey and Goodyer elms are dark, slightly leathery, broad or narrow, concave (and more or less folded, as mentioned above). The leaf serrations are said to differ in the number of secondary teeth (discernible at one-third of the distance from tip to base):

Cornish, 0–2 secondary teeth; Jersey, 1–3; Goodyer, 2–3.

Above and top centre **Cornish elms in Cornwall.** Top right **Hybrids in North Cornwall.** Bottom right **Flowering twigs of wheatley**

Melville has suggested a consistent geographical variation in breadth of leaf, for this and other native elms. Such 'harmonious' variations point to the nativeness of a species.

The shoots of *U. angustifolia* grow in beautiful curves, one overlapping another. The curved or sinuous twig is characteristic – not unique to the species, but it does have its own special curve and counter curve. Side shoots are short; twigs dark and hairy at first; and buds small, not pointed.

These characters are slightly less evident in the Jersey variety, which has rounder and flatter leaves and twigs more moderately curved.

U. angustifolia comes into leaf late (in May) and keeps its leaves late in the autumn. The dark, lustrous green is typical.

Cornish elms in Cornwall are blown into all sorts of sculptural and handsome shapes, but Jersey elms rarely vary from the conical shape. Very old trees become broad at the base. The variety *wheatleyi* is rarely distinguished by botanists from the Jersey elm. It is a well-mannered, narrow pyramidal tree with fine and erect branches, much planted in North London and Southport, among many other places. In London it is now sadly defoliated; in Southport it was a wonderful green in late October.

Perhaps the Wheatley and Jersey elms are botanically the same, and the cultivated variety is a clone selected from a multiple hybrid with *U. angustifolia* uppermost in its make-up. The Jersey elms of north-west France are reported to be hybrids of this sort. A similar variety was marketed early in this century, Elwes notes, as *U. campestris* var. *monumentalis*.

Cornish elms still form a noble and a vigorous element of Cornish scenery. Long may they remain – there was little disease there in 1976. Goodyer's elm, however, whatever its position in 1939 (*or* in 1624) cannot now be said to be thriving. There are more half-built boats, cosily named bungalows and alien conifers on the coastal plain of Hampshire than there are Cornish elms – and many, many more cars and lorries. The beech and oak country inland is, of course, carefully conserved, being part of our most ancient forest.

Plot elms finely grown and (right) **youthful in a back garden in the Trent Valley.** Opposite **Ancient Plot elms at Dry Doddington, near Grantham.** Foot **Plot leaves at Kew**

The Plot elm

The next native elm to be identified was not actually discovered by Dr Plot – his turned out to be a hybrid – but it was given his name. Melville says that it should be called the Plot elm, but not Plot's elm. Some people call it Druce's Plot's elm, but that is rather a mouthful. (Druce thought his elm was Plot's; but it should really be called Druce's elm, because it was not really Plot's at all. But Druce was a bit careless about how he published his find. Melville describes it as an idea which gradually formed in his [Druce's] mind.)

What happened was that Dr Plot, who published his *Natural History of Oxfordshire* in 1677, described a narrow-leaved elm planted as an avenue at Hanwell, near Banbury. Being also smooth it deserved the diagnosis *Ulmus folio angusto glabro*. It differed from the two other smooth-leaved elms in Gerard's *Herbal*. There the matter seems to have rested: Plot's elm became lost in the proliferations of *U. minor*, the small-leaved elm that never was, until Druce began to publish, in 1908, 1910 and 1911, various hints that he had found a small-leaved elm identical to the engraving in Plot's *Oxfordshire*. (The engraving, incidentally, makes it look like a limp piece of dog's mercury.) His trees were at Banbury and at Fineshade, in which locality they were called Lock's elm. Druce confused everyone by mentioning a number of dubious synonyms, and he was much criticized for being so vague.

Druce replied that he had a specimen actually collected by Dr Plot. Melville says that he was wrong in associating the Plot specimen with his Banbury trees, but that all the same he had recognized as a species one of our most distinctive native elms. Druce published photographs, and Melville was able to trace, in 1939, the very trees: in Banbury, in the garden of no. 17 West Bar Street. Specimens from these two trees are now in the Herbarium at Kew. The trees themselves were cut down by Banbury Corporation to make a car park. Dr Melville rushed to the scene just too late to save them – the very suckers were being torn from the ground as he arrived.

Melville's account in the *Journal of Botany*, August 1940, makes some interesting points on the distribution of Plot's elm, which was analysed afresh, since Druce's material was so confused.

'The main centre of distribution is in the Trent Valley around Newark on Trent, together with the neighbouring part of the Witham valley', wrote Melville. 'In this area the plume-like form of *U. plotii* is a common feature of the landscape. It is seldom found above 400 ft, preferring a deep soil with ample moisture . . . often found along river banks . . . its distribution is related to valley systems. . . .

'Hybrids with *U. glabra* are found, and within the limits of *U. plotii*'s distribution are more common than the species. The presence of the hybrid where the species is absent does not necessarily suggest a previous range of the species.'

U. plotii is scattered. Was it formerly more abundant? Dr Melville thinks it was: it produces good quality straight timber and was probably exterminated in many areas by timber merchants.

U. plotii Druce is a tall erect tree with an arching leading shoot when young. It has relatively few, short, almost horizontal branches from which slender, pendulous branchlets hang.

The leaves are small and appear smooth on both sides but are minutely scabrous above and hairy below. They are a dull green, paler below. The unique characteristic of the Plot elm is its ability to produce 'proliferating short shoots'. The terminal bud of most elms falls off: the Plot's just keeps on growing, in many cases producing a double handful of leaves, say eight instead of five, the newest leaves being rounder with less pronounced serrations. The

Plot elm taking over a hedge in Nottinghamshire, late summer

'main' leaves have fewer veins than other elms.

The bark of mature trees is dark grey, fairly regularly broken into rough vertical bars.

Unhappily, the plumes of *U. plotii* are no longer a common feature of the landscape of the Trent about Newark and the Witham above Lincoln. Elms are now few in these areas that were once the home of the Plot elm. A wartime shortage of wood, altered drainage levels, land clearance for power stations and machine farming have all combined into the familiar pattern of short-term efficiency and long-term degradation. Cultivated Jersey elms are a feature of the landscape of suburban Lincoln – another opportunity lost by a local council, for Plot's silhouette is no broader than Wheatley's, even though Plot might have looked a bit rustic amongst the semi-detached houses.

The few old Plot elms I did find in the Witham Valley were stag-headed – they had been short of water for years. A straggling group, now sadly diseased, shelters below Great Gonerby in a gulley too uneven for the tractor plough. A young Plot elm at Kew still survives in the autumn of 1977. This specimen has larger leaves than Melville's type tree. There is a small wood, with some very old pollards and a collection of younger trees variously lopped, straddling an ancient boundary by Mepal Church in Cambridgeshire. In this context the characteristic shape of the tree does not appear – more perceptive botanists may detect some hybridising with *U. carpinifolia*.

The Plot elm is a beautiful tree, and I should like to hear from readers who identify healthy specimens in their own areas.

Ulmus carpinifolia

Of the four elms in the second edition of Gerard's

Leaves of the tree opposite, U. carpinifolia

Herbal, 1633, three are identified as *U. procera*, *U. angustifolia* var. *goodyeri*, and *U. glabra* – Gerard's 'Witch Hasell or Broad leaved Elme'. The fourth was a smooth-leaved elm, *Ulmus folio glabro*, which he also called a 'Witch Elme', that being its local name in Essex. It was very plentiful between Romford and Stubbers, 'intermixed with the first kinde [*U. procera*] but easily discerned apart'. It was, in Essex, 'more desired for carts than the first'. Stubbers appears to have been at Ockenden, which doesn't sound elmy – but there *is* Elm Park on the way back to Romford. The description is obviously by Goodyer, who as a Hampshire man could hardly be expected to have explored the wilds of East Anglia.

This elm, with leaves smooth on both sides, must have been *U. carpinifolia*, a native of south and central Europe, North Africa, Asia Minor, east Kent and East Anglia.

Positive identification of this tree dates from 1773, by Gleditsch in his *Pflanzenverzeichniss*. Gleditsch worked at Jena, which is on the same

A smooth-leaved elm in an east Midland wheatfield

latitude as Dover. He provided no specimen of his tree type, but the Schleiden Herbarium collected a 'neotype', in the form of a variegated but wild specimen, from the Jena area in 1850.

Miller, in the 1768 *Gardener's Dictionary*, called it *U. glabra*. Linnaeus was no help – he called all European elms *U. campestris*. Some people think it would have been wise to leave it at that.

Moench (1794) struck out on a line of his own with *U. nitens* (shining), and then spoilt the effect with his variety *suberosa*. Loudon went almost back to Gerard in describing Miller's *U. glabra* as the wych elm, the most common elm in some parts of Essex.

Another name mentioned by Loudon was 'feathered elm' – which nearly stuck, for it appears in *Drawings of British Plants* by Stella Ross Craig (1970), and in Makins (1936).

U. nitens it remained until in 1946 Gleditsch, with his hornbeam-leaved elm, was rehabilitated by Melville in the *Journal of the Linnean Society* (Botany), vol. 53.

The smooth-leaved elm is described as an erect tree which can be as tall as the English elm but has a much broader spreading shape with a fairly open canopy. Branches are ascending and the trunk is somewhat sinuous; the crown may be a pyramid but is more often rounded by wind action; suckers are produced; and the twigs are sometimes cork-barked. The shoots from the main branches are long and pendulous: leaves 4 inches (9 cm) long to half that, smooth and shining, typically unequal at the base, regularly double toothed with one to three secondary teeth. The leaves are narrow, and an important recognition feature is the nearly straight lower half or third of the long side, which then makes a right angle bend to the leaf stalk. The latter is comparatively long and, like the bud, is lightly hairy. *U. carpinifolia* comes into leaf later than other elms, in late May.

‡ 3 *Vlmus folio latiſsimo ſcabro.*
Witch Haſell, or the broadeſt leaued Elme.

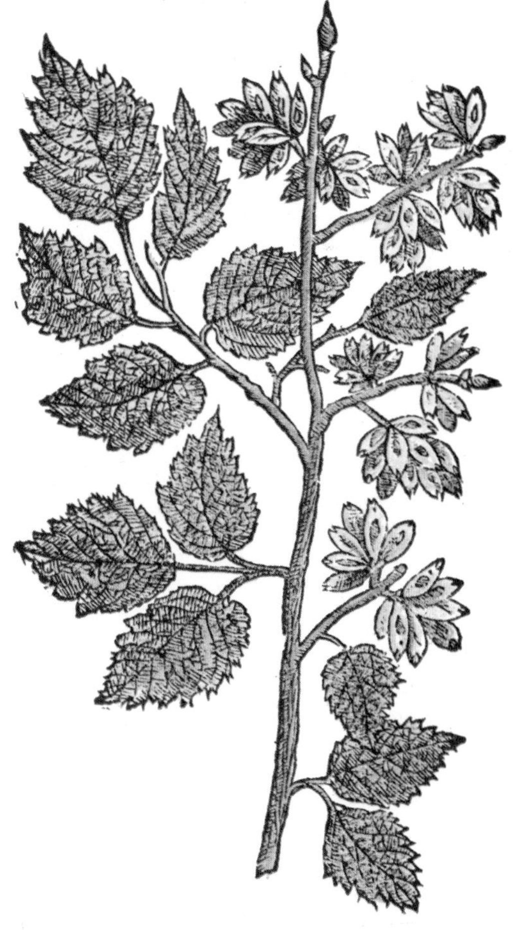

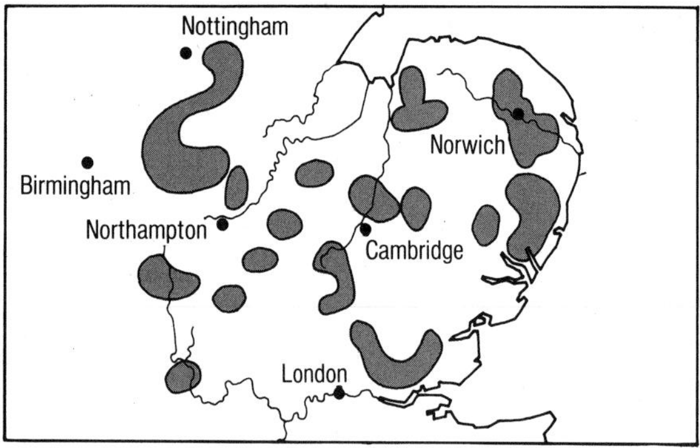

Above **foliage** and below **distribution of U. coritania**

Coritanian elm silhouette, Wicken Fen

The Coritanian elm

On 8 May 1947, Melville read a paper to the Linnean Society in which he identified a new species of English elm. He had found it in the area of the Midlands near Leicester once occupied by an ancient British tribe, the Coritanae. Melville had stumbled upon the new species while looking for specimens of *Ulmus elegantissima* Horwood, the Midland elm. Horwood could not accept some of the specimens as being true to his elm, which in fact later turned out to be a hybrid of *U. glabra* and *U. plotii*. Melville widened his investigations and came up with an endemic species containing three varieties distributed in a topocline: leaf shape varied by decreasing breadth and increasing asymmetry from south to north, from the Thames to the Trent. The species was distinctive, not by its spreading, open crown, but by the bright green, leathery texture and marked asymmetry of its leaves.

The species proved to occupy a much larger territory than had the ancient Coritanians: but the name is felicitous. Decent attention was given to the possibility of using existing names which had been given to specimens already in various herbaria. *U. nitens* var. *sowerbyi* Moss was one, and so was *U. nitens* var. *hunnybunii*; both of these are now regarded as hybrids. Some may regret the loss of these very English titles, one earthy, the other at least engaging, but I for one could not respect a tree called Sowerby's elm or Hunnybun's elm.

U. coritana is a spreading tree, up to 60 feet (18 m) high, with ascending branches; the crown is rather open. Leaves are leathery, bright green and shiny above, paler below, and smooth except for conspicuous axillary tufts. Leaf serrations are somewhat blunt with usually three or four subsidiary teeth. The leaves are strongly asymmetrical at the base, the midrib being curved towards the shorter side, and the longer side meeting the leaf stalk in a lobe. The leaf stalk is downy above.

The distribution map above is based on thirty-eight specimens gathered by Melville and his team before 1949.

A row of ancient pollards of *U. coritana* at Wicken Fen (ten miles north east of Cambridge) is probably unique. But Wicken Fen itself is unique, although it represents thousands of square miles of the old Fens: secretive, treacherous, yet highly productive of small timber.

These then are the four smooth-leaved elm species native to England. They have been here since about 5000 BC, when they occupied, as they do now, low, moist ground in the central and south-eastern part of the country. Except for *carpinifolia* they are peculiarly English and, it must be admitted, rather elusive and mysterious. There are even said to be other, even more obscure species in the lanes of Suffolk and Norfolk, awaiting their Melville to disentangle them from the multitudes of their hybrids.

They have been ignored and neglected, and they are ill-fated; lost causes such as the English love.

They lived through the Elm Decline (see Part 3), and, cinderellas of the tree population, they survived while cattle roamed the open land and cottagers cut all they could burn and stripped the bark in the longest pieces they could get. They served for piles and pipes, carts and coffins, keels and chair seats, but were hardly distinguished from their rougher brother, the common elm. Secretly they survived while their hybrids were enthusiastically planted in miles of avenues, or in plots for quick profits.

They outwitted the deviousness of botanists, who hid the elms' true natures amongst dubious identities and ambiguous nomenclature and, perhaps understandably, confused them with hybrids and varieties. They lived in grand obscurity in the age when elms were immemorial, and parks were kept up, and hedges were properly laid; while cast iron, steel and imported softwoods ensured their peace.

The native smooth-leaved elms even survived the elm disease epidemic of the late 1920s. And then they came suddenly into their inheritance as Melville sought them out, measured the subtle curves of their leaves, straightened out the miserable state of their titles and carefully filed their leaf certificates and places of abode. A late triumph, but a just reward was theirs in return for 7000 years of service and survival: they appeared as species 3, 4, 5 and 6 in the first edition of the *Flora of the British Isles*, 1952, having, apparently, also survived the war.

But even in the hour of triumph their old enemy, confusion, struck again. Clapham, Tutin and Warburg, or their printer, made mistakes with the diagrams. *Coritana* was labelled *stricta*, *carpinifolia* appeared as *plotii*, *plotti* was labelled *coritana* and *stricta* had to make do, not for the first time, with *carpinifolia*. *Diversifolia* got a good showing, with two diagrams labelled correctly.

But let us say no more, for a worse fate was to follow: in the second edition they were all lumped together as 'fig. 49: Short shoots of *Ulmus carpinifolia*'.

Now many botanists and writers on trees take the view that all the smooth-leaved elms are varieties of *U. carpinifolia*. This view is at least tenable, although no one has proved that *carpinifolia* and not some other elm is the parent species – it just happens to survive on the continent of Europe while the others do not. The position of Clapham, Tutin and Warburg is inconsistent, for their English smooth elms are not now described vaguely as subspecies or, definitely, as varieties.

Many genera have large numbers of species in *The British Flora*. Even ignoring such plants as the sedges or the umbellifers, with seventy-five and forty-three species respectively, the trees *Sorbus* (whitebeams) and *Salix* (willows) are treated with much more care. *Sorbus* has twenty species, only three of which are easily recognizable; *Salix* is awarded nineteen, plus seven subspecies and thirteen hybrids, not to mention a few varieties.

The English smooth-leaved elms are as distinct as some of the whitebeams (which are sometimes much more local in range) and as many of the willows, which hybridize as freely. But the latter have the advantage of a continental range.

The second edition (1962) of the *Flora* has this paragraph under *U. carpinifolia*: 'There appear to be many clones differing from one another in shape and habit. Various clones and clonal groups have been given specific rank but it is doubtful to what extent the discontinuities between these correspond to the discontinuities observable in the field.'

As the remaining plumes of the Plot elm nod in the winds of time, the *Flora* sounds a last discordant note: *U. plotii* Druce is now *U. carpinifolia* Gleditsch var. *plotii* (Druce) Tutin. And so, you merry botanists, how do you do?

In English, that is Tutin's Druce's Plot's hornbeam-leaved elm, of Gleditsch.

Pollard Coritanian elms, Wicken Fen

Hybrid Elms

Most of the elms growing in the fields are hybrids: some are cultivated varieties, some are natural crosses, and some are crosses between cultivated varieties and natural hybrids or varieties. The possible exception is *U. procera* infertile, or endemic where introduced varieties may flower later than the native tree. (These areas, unfortunately, are the ones most devastated by Dutch elm disease.)

Some of the important named hybrids and cultivars are listed below.

Small-leaved elm

Ulmus diversifolia, the lock elm or small-leaved elm, has already been mentioned – it was given specific status before it was realized to be a straight cross between *U. coritana* and *U. plotii* when these species were recognized. It may well be more common than either. Melville referred to it as the East Anglian elm. The key to its recognition is rather subtle: ten per cent of its foliage has symmetrical leaves – or, more accurately, about ten per cent of short shoots have symmetrical leaves, and some shoots have both types of leaf.

The colour of the leaves refers it to *U. plotii*, as does its straight stem, but the crown is spreading and open. The leaves are more or less rough above, but partake of the leathery texture of *U. coritana*, with rather regular, blunt serrations. The fine twigs branch at right angles (like the wych elm). Suckers are common, and the trunk sometimes divides into two straight stems.

The timber is tough and locks the saw. Actually, all English elm timber locks the saw when worked

Avenue of Huntingdon elms at Hidcote, Gloucestershire

wet with a handsaw, and it is apt to bind on mechanical saws.

U. diversifolia is found in hedgerows in Hertfordshire, Cambridgeshire, Suffolk and probably Essex and Norfolk.

Dutch elms

French elms mentioned by John Evelyn in 1664 were probably our Dutch elms, said to have originated in Belgium: *Ulmus × hollandica*. The synonym *U. major* refers to a particular clone, and there are many. Dutch elms (supposed to be absent from Holland) have the habit of *U. glabra* but a more open crown, and large leaves without the pronounced basal auricle of *U. glabra* and with larger, pinkish leaf stalks (over ⅜ inch). Suckers are numerous, and often corky. The bark is brown and varies in appearance from small flakes to shallow square plates.

U × hollandica var Klemmerii – still alive at Kew

Hollandica is really one of many crosses between *U. glabra* and *U. carpinifolia* which are the most common elms of Europe. Its distribution in England is largely south-central, and it is a frequent victim of Dutch elm disease.

I have not been able to satisfy myself that *U. × hollandica* exists as a recognizable type in the English countryside, though I have found many trees with the habit of *U. glabra* and leaves differing from the species – many, in the eastern Cotswolds, with leaves resembling those of *U. procera*. Such a hybrid is not supposed to happen naturally except when long, cold winters delay the flowering of *U. procera*. Perhaps it *can* happen at heights above 600 feet, or perhaps a widely introduced hybrid of this sort has become naturalized – especially, it seems, in the old area of Wychwood Forest.

Mitchell (1974) says *hollandica* is common from Cornwall to north Hampshire and from Wiltshire to the east Midlands, but rare elsewhere. The biggest known are at Marlfield, County Tipperary (although otherwise rare in Ireland), and at Saltram House, Plymouth. Hadfield (1957) says that *U. major* is by no means uncommon as a hedgerow and woodland tree in Britain.

Dutch elms grow vigorously. Opinions differ as to the value of their timber, from worthless (subject to heart shake) to at least the equal of *U. procera*. Often a deep brown-red, it has been used for fine panelling.

The Huntingdon elm

Not a feature of the municipality of Huntingdon with Godmanchester, this very distinct and elegant hybrid was so named (Hillingdon, about 1760) because it came from a nursery near Huntingdon. It is sometimes called the Chichester elm, for no good reason. It is *U. × hollandica* var. *vegeta* Lindley (1829) or Loudon (1838).

Huntingdon elms have the strongest growth of all elms in Britain. The usual form of the tree is narrow, a straight clean bole up to about 10 feet (3 m), and straight branches ascending at about 30° to the vertical with numerous sucker shoots on the branches.

The leaf is large, coarsely, variously and irregularly serrated, leathery and long-pointed. The base of the leaf is very oblique and the short side curves in to the first nerve – this is the key recognition feature, not present on every leaf. The leaf is long (0.5–0.75 inch – 12–19 mm) and pale brown.

Elwes and Henry (1913) did not recommend this tree for avenues. In spite of that, one of the most beautiful avenues, at Hidcote near Chipping Campden, is of Huntingdon elms. Here they form a large nave with the exact geometry of the Early English arch, but without, of course, the rigidity of stone and with the advantage of a soft green carpet underfoot*. The timber is red.

* But they were nearly all dead in 1977.

Left Huntingdon elm in Highgate Crescent, London N5
Right A mysterious graft in Regent's Park

Left **Magnificent hybrids in East Suffolk – leaf resembling Huntingdon.** Top **Near Banbury form resembling Huntingdon, leaf close to *U. procera*.** Right **Enormous leaf of *U × hollandica wentworthii***

There are many other hybrids of *U. glabra* and *U. carpinifolia*, besides *hollandica* and *vegeta*. Trees labelled 'Dutch elm' at Kew have long, rather narrow leaves of leathery texture – some are triple hybrids: *U. carpinifolia* × *glabra* × *plotii*.

A sort of Dutch elm actually raised in Holland is Commelin, thought to be resistant to Dutch elm disease, but not so. A tree with enormous leaves (7 inch (18 cm), but they look bigger) at Kew is called the Wentworth weeping elm. Several magnificent dark trees around the Inner Circle at Regent's Park have similar leaves, slightly smaller on less pendulous shoots.

The Midland elm

This was *Ulmus* × *elegantissima* Horwood, now known to be *U. glabra* × *U. plotii*. It is elusive, both in the textbooks and in the field.

Some Other Elms

The native American elm is a large and beautiful spreading tree – or was, before the 1960s strain of Dutch elm disease swept across all North America. It is the American white elm, called white because of its timber; but the bark is sometimes pale grey. The small purple flowers, in March, are pendulous. The fruit has hairy margins with an oval seed in the centre. The leaves are medium sized, rough to the touch, deeply double-toothed, with rather close, parallel veins.

Ulmus americana, readers will be relieved to learn, has never had any other name since Linnaeus's classification (1753), before which few botanists can have seen it. Though described as very variable in habit, it has avoided the botanical splitters, except for some early synonyms, *mollifolia, pendula, alba,* and *floridana*, which we can ignore.

Above **The American elm.** Opposite top left **U. villosa.** Top right **U. parvifolia, the Chinese elm.** Bottom left **Leaves of the slippery elm.** Bottom right **Trunk of U. laevis.**

Specimens surviving from the original American forests were described as magnificent, with clean cylindrical trunks 4 or 5 feet in diameter up to 60 or 70 feet, where they divided into three or four great limbs which then diverged on all sides with long, pendulous branches which seemed to float lightly in the air.

'No tree has attracted more attention among American writers or is more dear to the native of New England than the American elm', wrote Elwes, 'which is a conspicuous ornament and the favourite shade tree in the older cities, and has a literature of its own.' Elwes described the various forms of the tree:

Vase, umbrella, plume, drooping and oak-form

The timber, called white, grey, water or swamp elm, is one of the strongest of elm woods, good for

Leaves of the Canadian rock elm, U. thomasii

tool handles and wheel hubs, which used to be exported to England ready turned and morticed.

The tree is a native of eastern North America including part of Canada.

Two other well-known American elms are the Canadian rock elm, for its rock-hard timber, and the slippery elm.

The Canadian rock, cliff, cork or Hickory elm is *U. thomasii* Sargent, 1902 (*U. racemosa* Thomas, 1831): French, *orme à grappe*. It is remarkable for its heavy, hard, compact, very strong, tough and elastic timber, to quote the words of Boulger, the author of *Wood*, 1902. The buds are hairy and sharp pointed, the leaves rather ordinary, the twigs usually corky. The small specimen I have seen has numerous burrs with sucker shoots, but this is almost certainly untypical; the tree is unsuitable for our climate. The height of the fully grown tree is given as 100 feet. The rock elm flowers in drooping racemes in April. The natural range is from Quebec to Florida, over the whole of the eastern half of the continent – so perhaps Canadian is not a good name.

It is a very slow-growing tree. Elwes counted 250 annual rings in a 9½ inch log (which means that each ring was about 0.04 inch apart). One tree was reported to have grown only 1 inch in diameter in fourteen years.

The slippery elm is best known for its use as an old-fashioned remedy for coughs, tummy pains, typhoid fever, toothache – practically everything. The inner, mucilaginous bark of ten-year-old wood is used powdered and mixed with water – or even with milk and eggs for gruel or infant food: it is as nutritious as oatmeal, but tastes insipid and can be flavoured with cinnamon, lemon peel or nutmeg. Patent foods are made from it, especially from the slippery elms of South Michigan. The Indians used it as a poultice for wounds, boils and burns. The timber, American red elm, moose elm or slippery elm, *orme gras*, is important – or was – and the tree grows vigorously all over eastern America. It is the American counterpart of the wych elm, growing on steep banks and hills. The wood splits easily and is coarse grained, but it was used for the ribs of boats, for railway bridges and for fences – an important material in the winning of the West.

The tree is *U. fulva* A. Michaux (1803) or *U. rubra* F. A. Michaux (1813). To Linnaeus it was just another form of *U. americana*. It is a relatively small, solitary tree. The twigs are tinged with orange, and the large buds are downy or even long-haired. The leaf is large, flat and floppy, with pronounced triangular teeth, and it is a clear yellow in autumn. The flowers are erect, in March.

The wahoo, or winged elm, *U. alata* Michaux, has a range similar to that of the other American elms, but tending to the south: wahoo is the Indian name. Not a very large tree, it is named for its cork-winged twigs. The leaves, and the fruit, are narrow, the flowers drooping on slender stalks. The leaves are only 1½ inches long and have hairy margins. The timber, though usually small in section, is heavier and closer grained than that of *U. americana*, and not so dense as that of the rock elm.

The European white elm, *U. laevis* Pallas (1784), is a native of central Europe and West Asia (Russia up to latitude 63). Like the American white elm, it has always been distinctly identified by botanists, though there was a certain amount of indecision about its name. *Laevis* means smooth. It was named *U. pedunculata* by Fougeroy, also in 1784, because of its pendulous flowers, and to

Loudon it was *effusa* – he called it the spreading-branched elm.

The leaves are rather like those of the American white elm but usually smaller, and, in specimens I have gathered, more distinctly oblique at the base, the midrib curving into the leaf stalk and the short side of the leaf as much as three nerves short. There may be sixteen or seventeen veins on the long side, which is more than any other except *U. glabra*; serrations are complex, curved and sharp. The buds are distinctive, sharp pointed and a nice orange-brown.

Mitchell (1974) describes the tree as having an untidy crown, arched branches carrying fine sprouts and burrs, and dull grey-brown bark with a wide, shallow network of broad smooth ridges. But the bark of young trees is smooth, hence *U. laevis*.

This white elm, says Elwes, is especially abundant near Brandenberg in damp deciduous woods, and in north-east France in oakwoods, never in quantity. It grows in the Crimea and the Caucasus, and is planted along railways in Russia to protect them from snowdrifts. Although supposed to be hardy, it is rare in Britain, where it flowers in March and comes into leaf early. The timber is hard and durable, and is used in Russia for the same purposes as the English elm in England.

A beautiful elm of the western Himalayas is *U. villosa* (soft haired) with silver bark, ringed like a cherry on the branches, but heavily and vertically striated on the trunk. The leaves are somewhat delicate, neither tough nor smooth, and very lightly pubescent; with their long, thin leaf stalks on fine twigs, they look a bit like young ash leaves as you look up into the tree.

The form is upright and open, branching irregularly. This lovely silver elm would be a great asset as a street tree. The specimens at Kew seem to be healthy, and one cannot imagine the *Scolytus* bark-beetle recognizing this tree as a suitable home.

The Siberian elm is *U. pumila* Linnaeus (dwarfish), with rugged, ridged and irregular bark, a compact crown and small, irregularly toothed leaves on long stalks; it is nearly evergreen, but the specimens at Kew have died.

The Chinese elm, *U. parvifolia*, has very small, shiny, dark green leaves which remain thus until late November. A completely evergreen variety, *sempervirens*, originating in south China, is cultivated in Florida. The crown of this tree is dense, giving a deep shade, but the one at Kew is dead.

This and *U. serotina*, the red elm of the American South, are September-flowering elms. *U. serotina* ('late in the year') has $2\frac{1}{2}$ inch leaves, brilliant red in autumn on silver twigs, which can be corky. The flowers are pendulous, and the fruit hairy.

The rough bark of the Siberian elm, U. pumila. This species was thought to be disease-resistant, and the small-leaved variety aborea was much planted in English gardens

U. japonica, unlike other *japonicas*, is a large tree. The leaves are large and broad. Reports differ on its actual habit, and I have not been able to visit Japan to see it. Everett (1969) says 'broad-headed, up to 110 feet high and like a Scotch elm'. Makins (1936) says 70 feet, and like an English elm. It flowers in March.

Kamui Fuchi, the chief goddess of the Ainu aborigines of north Japan, was born from an elm impregnated by the Possessor of the Heavens. The elaborate rituals of this very tree-oriented tribe make use of every sort of wood, with preference for laburnum and willow for their intricate and elegant prayer sticks. The Ainu's elm is used for thatching (the outer bark) and cloth (the inner bark) and the rest, including the roots, for fuel, thus returning the tree to the Possessor of the Heavens.

The Genus Zelkova

The genus contains about six species of trees resembling elms, the fruit being a nut instead of the winged samara of the elms.

'If ever a tree deserved a good common name,' writes Hugh Johnson, 'this is it'. Loudon knew it as *Planera richardi* Michaux or 'Richard's Planera' – Planer was an otherwise forgotten professor of botany. The name does not sound right for this tree, with its upward-springing form and pretty leaves. It is usually called the Caucasian elm, and this is confusing.

Zelkova carpinifolia is a native of the Caucasus and is perfectly hardy in England, if slow-growing at first. The shape of the leaves can be seen from the picture, unmistakeable, and not much like hornbeam. The habit of the tree is variable and may be somewhat reminiscent of the hornbeam; the bark is beech-like, dark or light grey according to surroundings. Small flakes reveal bright orange underneath.

The most photographed specimen in England, in a sunny Devonshire paddock, has a quite extraordinary fluted bole surmounted by a great explosion of small branches. I think it was polled, one hard winter long ago. Trees in London parks are certainly many-branched, but have taller trunks dividing gradually into heavy limbs.

Z. *serrata* Makino, the *keaki* tree of Japan, has simple, sharp leaf serrations and the leaves are not particularly asymmetrical. They are placed regularly on the twig, which is hairy. The branches are more spreading than those of the Caucasian elm.

Z. *sinica* Schneider has only a few small sharp teeth on the leaves, which are entire (plain) at the base. The bark of this Chinese *Zelkova* is very richly patterned as if with a free inlay of pinkish gold and grey over the short trunk and the well-shaped branching limbs.

Zelkovas are being planted as street trees instead of the elms. They are sometimes grafted onto elm stocks, which helps to raise the branching height. All the zelkovas turn to rich browns, pinks and oranges in autumn, and hold their leaves late. The timber is fine and durable.

There is a cut-leaved variety of unknown origin with very sharply toothed leaves and a fierce name. *Zelkova verschaffeltii* Nicholson.

The two larger photographs are of a fine tree at Pitt Farm, Chudleigh, Devon: Z. carpinifolia

Above left Trunk of the Chinese zelkova, Z. sinica; leaves of Z. carpinifolia and (right) Z. serrata

Part 3
The Elm in the History of the British Isles

Prehistory

Above **Glacial moraine lakes on Rannoch Moor – woodland restricted to ungrazed islands.** Below **Sea buckthorn, Hippophae rhamnoides.** Right **British woodland – here of holly and oak – in a frost free, damp climate**

About 10 000 years ago our present postglacial, the Flandrian, followed the end of the Devensian (usually called the Weichselian) glacial age, marked by a sudden rise in temperature. This is shown, in the fossil pollen analysed by paleobotanists, by the rapid spread northwards of the tree birches (*Betula pendula* and *B. pubescens?*) over the late-glacial tundra of grass, sub-arctic flowers and mosses which covered lowland Britain and most of what we now call the North Sea. The tree flora of the tundra consisted of herbaceous willows and birch; and sea buckthorn – a plant which now divides its habitat, strangely, between the Swiss Alps and the sea shores of Essex and Lancashire. There were wide heaths of crowberry, such as now, apparently, exist in northern Scandinavia. In favoured places were small woods of birch and aspen. Low juniper scrub covered higher ground and must occasionally have found hillsides suitable for the formation of taller juniper

woods such as that which survives, in exquisite formal perfection, at Tynron, Dumfriesshire.

Ice still covered Scotland, and great Cumbrian, Cambrian and Pennine glaciers produced torrents, draining into what may have been a great cold swamp surrounding the plateau of the Dogger Bank. Here the Thames, the Humber and other great streams unnamed met the Rhine.

Massive Irish deer or elk roamed the open land and ate the small trees. Mountain hares, perhaps, lived in the grassland, and packs of wolves hunted the pinewoods of what are now the southern North Sea, Norfolk, Hertfordshire, Hampshire and the Channel, across to Normandy and Brittany.

The great wave of birch and pine forest spread quickly north and west, preceded or accompanied by abundant juniper. A few miles walk from the Glen More Caravan Park, under the Cairngorm, will take you into woodland of this gorgeous pattern: climb perhaps 1000 feet (300 m) up the north-west side of Carn Eilrig, and you may see spread before you the chequered light and dark green, broken by the gashes of snow-swelled streams, that must have been typical of much of England, eighty centuries ago.

Grazing animals probably kept a good deal of the grass and scrub free of pines, as they do, far too much, today. Many pine forests, also, are known to have been relatively short lived. In some places they were quickly superseded by oakwoods, in the succession which is still effective in minute areas of the Western Highlands. The oaks then succumbed to rising mires, which in turn were drained to large areas of peat, open to the possible regrowth of pine. The picture of the dominance of pine may be somewhat distorted by the density of the pollen showered from these trees and the nature of the ground which happened to be that most likely to preserve it.

Almost immediately after the sudden spread of birch, and sometime before that of pine, came a virtual explosion of hazel – which became, in Ireland, as much as seventeen times more important than any other tree, and in England often four times as important.

Here our picture of the ancient forest becomes a little less clear. We have established in our minds a cover of pine with birch and juniper and, perhaps, heather or bilberry. Now we must envisage a great ocean of hazel scrub. As an undershrub to pine it is unfamiliar, although apparently it does so co-exist in the Baltic island of Gotland. Soil and climate in the Boreal Period were both different from anything we are now familiar with. In the western part of Britain hazel must have formed considerable woods, in the absence of competition from taller trees, on the new soil of evaporating moraine lakes. *Corylus avellana* is by no means just the delicate-looking shrub of oakwoods, hung with lambs' tails in early spring: it can make a very sturdy tree. Some neglected hazel coppices can now give a good impression of the character of hazel-dominated woodland. A dense canopy is formed at about 25 feet (7.5 m), shading all but early spring flowers. But only in the Burren of County Clare can we currently see the hazel in its role of pioneer woodland species.

At about the same time as the sudden expansion of hazel, there appeared in Britain another pioneer, Old Stone Age man. Nuts were an important part of the diet of this explorer and hunter, and he carried his supplies with him.

We have now set the scene, about 8000 years ago, for the entry of oak, elm and alder into the composition of the native forest. Oak and elm expanded dramatically at about the same time, at a period of considerable increase in warmth and moisture. Lime and alder followed. All the trees varied in distribution according to soil and altitude. These colonizing trees, including elm, were not unknown in Britain before 6000 BC; they were just very rare and struggling in pockets not affected by sheet ice, and confined to the south.

As the climate improved, over some 3000 years, to reach what must have been the ideal for north-temperate forest, the mixed oak–elm–lime–alder woodland grew to a magnificent climax. Eventually, at a point in the Atlantic Period, trees of some kind occupied virtually every piece of soil, to the almost complete exclusion of open grassland or heath.

Regional variations appear to have been much as they are today, but greatly simplified, as if in a boldly coloured diagram: for each tree species was without competition from others within quite large areas of soil, position and precipitation. Birch and pine in Scotland dominated the landscape, and grew up to a tree line at least 1000 feet above the present (about) 2000 feet. On loch sides and river banks were miles of alder. *Alnus* pollen in many northern and western deposits is up to sixty per cent of the total tree pollen – no doubt partly because its habitat favours pollen accumulation and preservation.

In the lowlands, great tracts of oak, lime and elm also grew to the almost complete elimination of other plants. Because oak usually predominates, this type of woodland is often referred to as oakwood. It is true that oaks grow slowly and tend to a fairly open pattern, with sporadic colonization by other trees and shrubs. Today the pedunculate oaks, *Quercus robur* (or *pedunculata*), and the sessile oak, *Q. petraea* (or *sessiliflora*), are also highly tolerant and will invade territories apparently better suited to other trees. All the same, it is not possible to assume that the 'mixed oakwood' of the Atlantic Period corresponded to the mixed woodland of today, which is the result of 5000 years of continuous interference by man, and the interplay of various ecological networks, including alterations of the soil, naturalization of new species of plants and animals, and the adaptation of existing species to a series of new habitats. Until some courageous worker attempts a survey of total post-glacial ecology, we can only speculate. It is at least likely that the woodland of the climatic optimum took the form of consistent bands and patches of single species of tree, with or without an understorey of, say, holly or hazel.

There is no living example of the climax forest. The nearest approach is in the oak and holly woods near Lake Killarney, where the moist, frost-free climate allows the vigorous growth of brilliant green mosses on trees and stones alike, the branches of the oaks are hung with lichen and epiphytic ferns, and the trunks are covered with ivy. The cleanness of the air is reflected in the silver of the holly trunks (see the opening picture).

Nearly all our native species are capable of forming woods – that is, in certain circumstances they become dominant to the exclusion of other

The dwarf birch, Betulanana, now found only on high moorland. Larger picture **Pure juniper wood at Tynron, Dumfries**

Top left **Wood of wych elm in the Scottish border country.** Bottom left **Elm wood spreading down the cliff at Cadgwith, Cornwall.** Top right **Carleon Cove, Cornwall. A wood of Cornish elms.** Bottom right **English elms forming a wood in protected Chiltern grassland**

trees. Pinewoods, birchwoods and oakwoods are familiar; hazel we can imagine. Small-leaved lime, *Tilia cordata*, we know can form woods, but examples are rare. Alder and willow are not perhaps good examples of single-species woods, but narrow alderwoods do still climb the banks of Welsh rivers and patches of sallow are quite common wherever agriculture allows a space on wet land. Ashwoods and beechwoods were not elements of the early Atlantic forest. Only elm remains as an important tree quite unfamiliar as a woodland type.

Wych elm, it is true, grows in patches along riversides in the north, and these patches would become woods if the surrounding meadows were not used. But we have no reason to assume (as is often assumed) that all the pollen records of elm apply to *U. glabra*. The wych elm is the only native elm in Ireland and probably in Scotland. This is consistent with its being the earliest elm colonizer of Britain. In a climate not less but much more favourable to the elms, there is every reason to suppose that all our present-day species were present and thriving, and only just beginning to suffer from the competition of their own hybrids.

I believe we should think of Atlantic elmwoods, and in the terms described by Ovington (*Woodlands*, 1965), who mentions the light level measured in a German elmwood as only one-hundredth of that above the canopy. In a wood as dense as this, other plant species are restricted to parasites, saprophytes and a few early flowering herbs. Invasion by other trees would be slow indeed, since the collapse of individuals through age would leave gaps more likely to be filled by elm seedlings than any others; a similar situation prevails in beechwoods today.

The covering of the land by trees very much reduced the habitat for large game, and the Old Stone Age hunters tended to be driven to the coasts – westwards and south-westwards, where they adapted to a diet of mainly shellfish, and northwards, where they became more or less settled on high land. A new invasion, or series of invasions, occurred: of Mesolithic men. Their tools were of flint and other stones, of bone and of antler. Barbed spears, used for freshwater fishing, are typical of the Middle Stone Age. One was found under 37 feet (11 m) of sea water off the Norfolk coast. Small sharp flints were fixed in handles of wood or bone. Nets were made of bark twine, probably from the inner bark of lime trees.

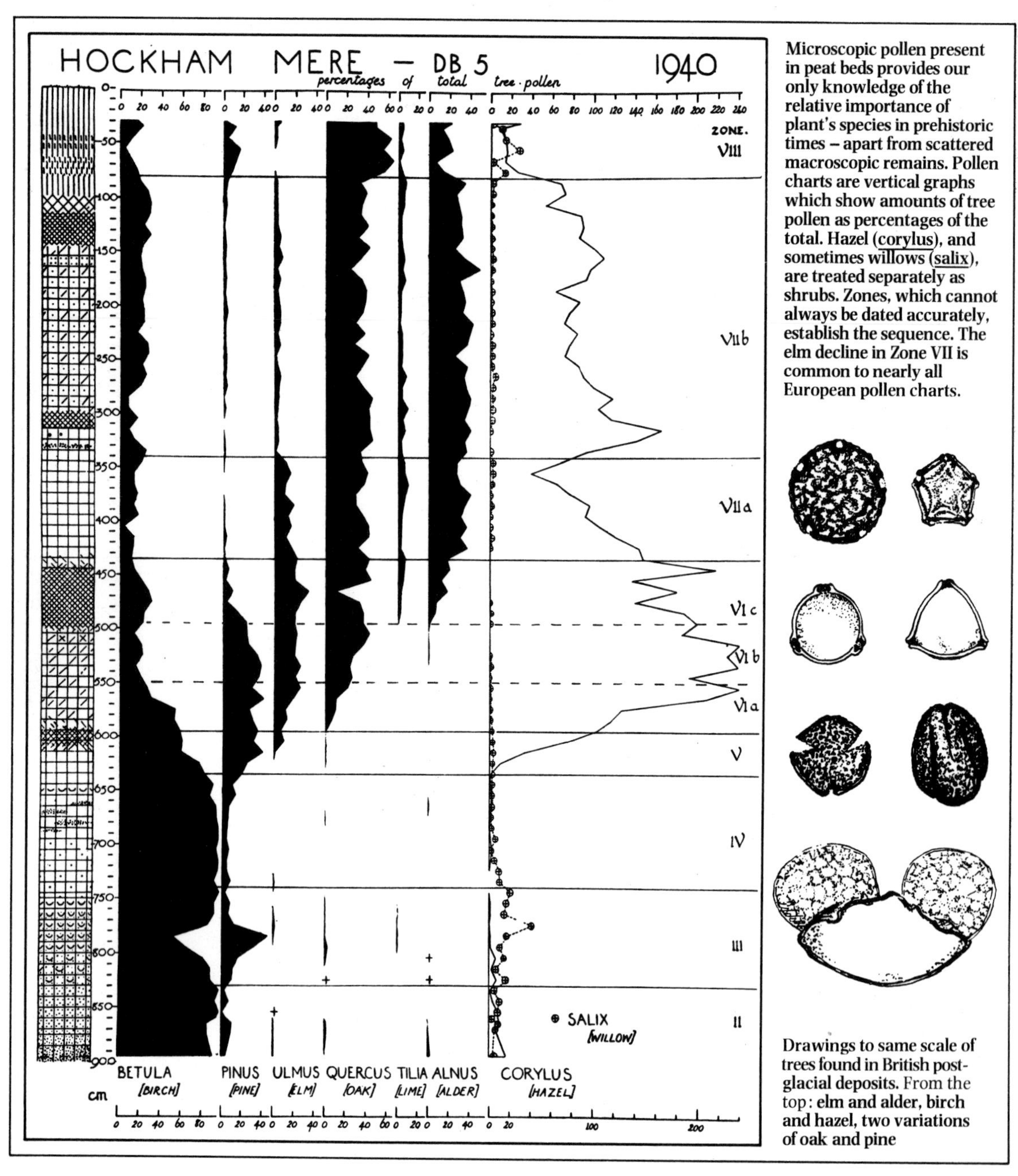

Microscopic pollen present in peat beds provides our only knowledge of the relative importance of plant's species in prehistoric times – apart from scattered macroscopic remains. Pollen charts are vertical graphs which show amounts of tree pollen as percentages of the total. Hazel (corylus), and sometimes willows (salix), are treated separately as shrubs. Zones, which cannot always be dated accurately, establish the sequence. The elm decline in Zone VII is common to nearly all European pollen charts.

Drawings to same scale of trees found in British post-glacial deposits. From the top: elm and alder, birch and hazel, two variations of oak and pine

The Elm Decline

For twenty centuries at least, the British forest grew luxuriantly and continuously from sea level to 3000 feet. Perhaps some downs and wolds were kept to grass by the last herds of deer – we do not know: no record is preserved in chalk.

The forest also continued below our present sea level, eastwards into Europe as far north as the Dogger Bank, off the coast of Yorkshire and Durham. The greater part of the sea bottom of the southern North Sea, even allowing for sand, is less than 100 feet below East Anglia.

Elms, which do not grow above about 400 feet in our climate, were probably restricted to 600 feet in the Atlantic Period, but the extra 100 feet of what we may call the Rhine Delta must have provided an enormous extension of ideal elm-habitat – and also indicates the likeliest invasion route. We may visualize for, say, *Ulmus carpinifolia* (without a shred of evidence, of course) a comparatively vast distribution area of which the upper Elbe and East Anglia are only the extreme tips. The same may well have applied to other elm species endemic to England – they are merely obliterated in Europe.

U. procera with its vegetative habit of reproduction and liking for open habitats is more difficult to speculate about. It is, however, not infertile, and in conditions of greater winter warmth and with moist soil ready for germination it may well have behaved like the other elms, but with a preference for well-drained grassy hillocks. I know of a spreading, semi-natural group of *U. procera* in protected grassland on the Chiltern escarpment (where chalk is obviously not, at that point, immediately on the surface).

The supposed incapacity of *U. procera* to reproduce from seed may simply be a result of our cold spring climate. It is probably at the extreme edge of its range, in the greater part of which it is extinct. Since the other elms hybridize freely, one might expect to find hybrids of *U. procera* somewhere in Europe. This does not seem to be the case. However, its orginal range may be now under the North Sea. We shall never know; pollen recovered from the North Sea moorlog (undersea peat) is said to be almost unrecognizable, and the possibility of macroscopic remains, even of a timber renowned for its long life under water, seems remote after several thousand years.

The maximum reduction of the sea level is supposed to have been up to 1000 feet (300 m) at the height of glaciation. During the post-glacial,

Ancient tree stumps recovered from peat, Connemara

replacement of sea water increased to a rate of 5 feet per century in the Atlantic Period. It is, apparently, rather difficult to estimate the extent of exposed land at any given time because the rise caused by the melting ice is to some extent and in different places offset by the general rise and fall of the earth's crust. This is to say, while the sea was filling up with melted ice and flooding the Channel and the North Sea (and the southern Baltic and the Danish seas) the land in northern Britain and Ireland was reacting to the removal of the weight of ice by rising by 25 feet or more. This is proved by the existence of numerous raised beaches. Slight eustatic rise still continues now, but the main movement seems to have been completed by 3500 BC, and the modern north-west coastline became more or less fixed as the rise in the sea level slowed.

The coasts of the southern North Sea are, as everyone knows, much closer to sea level. Here a slight down-warping or subsidence balanced the rise in the north-west and the coastlines were still fluctuating for about 1000 years around 3000 BC, the date of the Elm Decline.

The Elm Decline is the phenomenon which marks the beginning of a new, warm and dry 'continental' phase in the post-glacial climate. Large quantities of elm and lime pollen were seen as indicators of the climate improvement around 5500 BC. Lessening amounts therefore indicated a worsening climate around 3000 BC – a somewhat circular argument.

There is in fact no proof that the climate changed at all dramatically. But the amounts of elm pollen were dramatically reduced, and consistently so over a wide area. All tree pollen was

British woodland in the present Interglacial

As the ice retreats, pioneer species colonise the new soil. Temperature and rainfall reach a climatic optimum in the Atlantic period, when forest covers nearly all the land. Then complex factors of land-use, soil-change and climate work together to reduce the forest to its present 7 per cent of the total land area

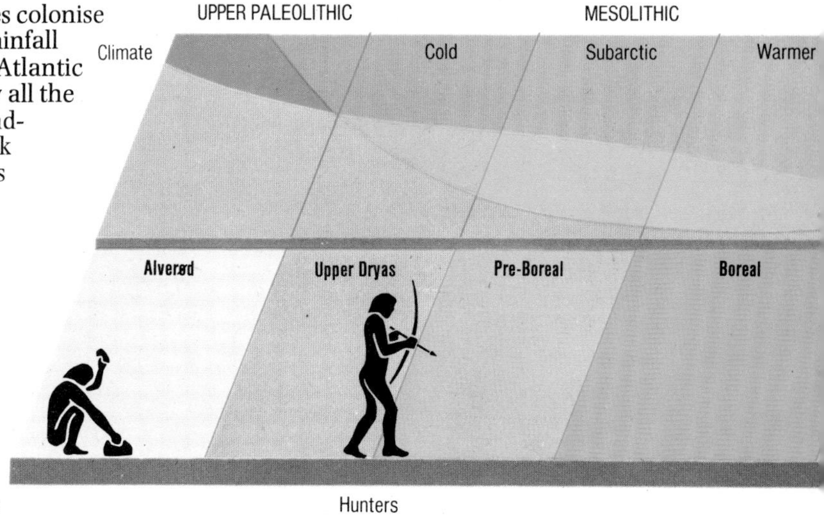

UPPER PALEOLITHIC
MESOLITHIC
Climate
Cold
Subarctic
Warmer
Alverød
Upper Dryas
Pre-Boreal
Boreal
Man
Hunters

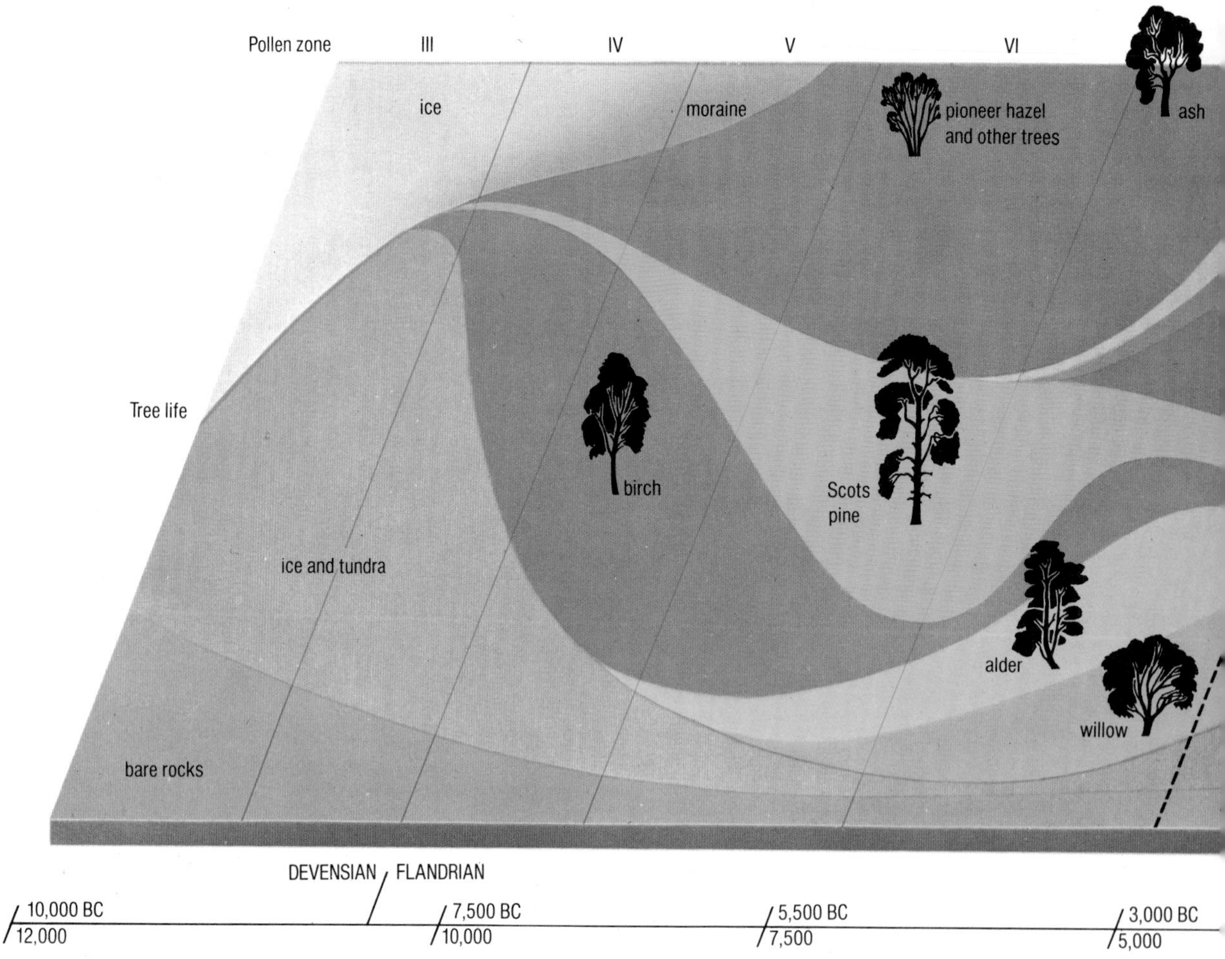

Pollen zone
III
IV
V
VI
ice
moraine
pioneer hazel and other trees
ash
Tree life
birch
Scots pine
ice and tundra
alder
willow
bare rocks

DEVENSIAN / FLANDRIAN
10,000 BC / 12,000
7,500 BC / 10,000
5,500 BC / 7,500
3,000 BC / 5,000

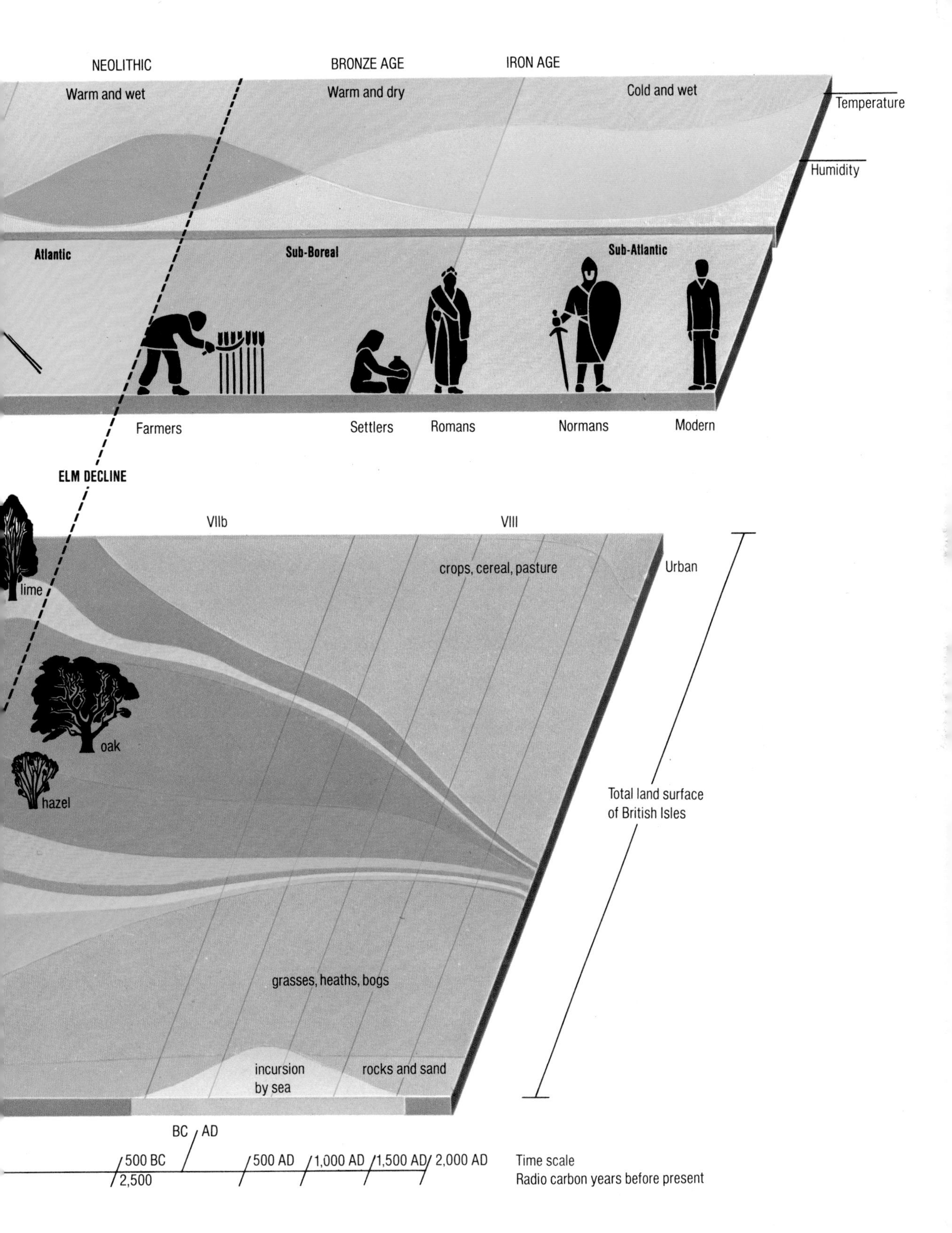
NEOLITHIC
BRONZE AGE
IRON AGE
Warm and wet
Warm and dry
Cold and wet
Temperature
Humidity
Atlantic
Sub-Boreal
Sub-Atlantic
Farmers
Settlers
Romans
Normans
Modern
ELM DECLINE
VIIb
VIII
lime
oak
hazel
crops, cereal, pasture
Urban
Total land surface
of British Isles
grasses, heaths, bogs
incursion
by sea
rocks and sand
BC / AD
500 BC
2,500
500 AD
1,000 AD
1,500 AD
2,000 AD
Time scale
Radio carbon years before present

reduced, but most trees somewhat recovered – elm did not. Many interesting theories have been tried to explain the decline of elm:

a The Flood.
b The stripping of the bark for fibre weaving by late Stone Age man.
c The use of the foliage for fodder, for beasts penned in the forest.
d Selective clearing of elm land for agriculture.
e Dutch elm disease.
f Climate becoming unsuitable for some elm species.
g Soil modification.

Well, the Semites, who were named after Shem, the son of Noah, began to settle in Canaan in about 3000 BC – the Hebrews were later immigrants. Biblical history does not usually take account of the local rise in sea level which must have caused the flooding of many coastal plains all over the northern hemisphere. Thunderstorms of Biblical intensity may well have occurred in a period which was a climatic optimum in northern Europe – and would go down in 'history' as the cause of the floods.

The records of peat on the East Anglian and Dutch coasts are confusing. Godwin (1975) admits that the question of the dates of these coastlines is open to argument: 'The displacements of deposits formed at sea level 6500 years ago in southern Britain . . . shews a depression in the southern North Sea of the order of six metres . . . but with such small movements as this, large tidal ranges, effects of compaction of sediments and the deduction of height or formation in relation to contemporary sea level, there is room for argument and reappraisal.'

Just possibly, temporary flooding of low land might have wiped out a lot of elms. But this does not quite explain the Elm Decline in Ireland, where the elm was often the co-dominant species with oak or hazel.

Neolithic man

Theories *b*, *c* and *d* above can be dealt with together. The arrival in northern Europe of people with the skills of Mediterranean farming began the process of forest clearance which continued until the seventeenth century. Neolithic cultures were various, but had in common the sort of organization that built up a flint-mining industry with 30 foot shafts at Grimes Graves (Thetford, Norfolk), coppiced hazel to make trackways on the Somerset Levels, and cleared most of Breckland's trees for good. In Scandinavia there are records of a clearing-and-burning method of land conversion, and there also survive bundles of elm fodder. In Britain there is no evidence that these methods were used, but elm fodder is traditional, down to recent times.

There is no particular evidence for the widespread use of elm bark for making weaving fibres, but the fact that lime as well as elm suffered a quick and apparently selective decline makes this theory attractive, for barking of these trees is very destructive. Lime bast (inner bark) is still used in some parts of the world, and Welsh thatchers used wych elm bark as ties until comparatively recently. And flour made from elm bark was part of the Lapplander's diet up to the nineteenth century. Really vigorous use of the bark, combined with the grazing of seedlings and saplings by domestic animals, would certainly allow a comparatively tiny population to destroy most of the elms and limes.

Whatever method was used to clear the land – and it has been proved that three men with stone axes could cut down 600 square yards of forest in four hours – the grazing of cattle is the sure way to prevent regeneration, usually of all trees except the hawthorn and blackthorn.

At various sites in Northern Ireland the Elm Decline was followed by the introduction of grassland, and Godwin states that there is conclusive evidence that the elm decline was 'achieved' in as little as thirty years, by clearing for cattle farming. There is also some evidence that the fertile areas sought out by Late Stone Age farmer-prospectors were those that contained abundant elm. Is it too daring to suggest that stands of pure elm were picked out as indicating fertile land? By almost magical coincidence the tree provides a store of fodder for the first winter, and a source of fibre which for all we know was essential to Neolithic husbandry – for thatch perhaps, for there cannot have been straw, and reeds might be miles away.

Even if clearances were not selective, the best of modern knowledge points to a human cause for the Elm Decline. It seems that no frost indication, such as the decline of ivy and mistletoe which was noticed in Scandinavian sites, has been traced here. Ivy in any case is also a fodder tree.

Forest clearances were followed by an increase in grass, plantains and nettles; early forms of the cereals wheat and barley are found.

An increase of ash pollen in some sites follows or coincides with the Elm Decline. Amounts are small. It has been said, with an air of finality, that

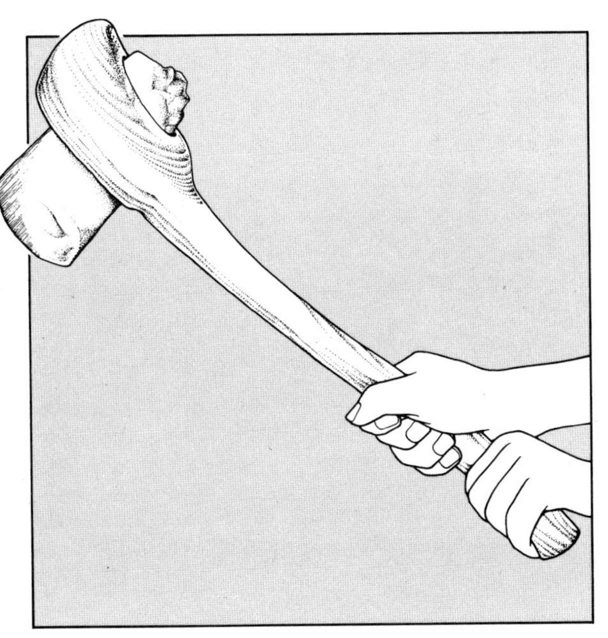

Using stone axes, men cleared the trees to grow crops

since ash is just as warmth demanding as elm this is an argument against a cooler climate. But ash will flourish on exposed high rock, particularly limestone, where only wych elm could possibly compete – and even then would cling to the hollows. Ash was probably one of the trees which benefited from clearance of dense woodland – and, as always later, it was almost certainly cherished for its usefulness.

Still, the climate did get cooler and drier. Bogs became blanket bogs, and some plants, like rock roses from Yorkshire, retreated never to return. But there does seem to be an inconvenient shortage of important species that flourished here in the benign Atlantic, left fossil records and then ceased to grow for lack of heat or moisture. The strawberry tree, for instance, which still fights a losing battle in the wild above the lakes of Killarney, has left doubtful fossil remains locally – and not at all in Devon or Cornwall where some Lusitanian flora does survive. Godwin is able to point only to the obscure greater naiad, a submerged waterweed now found in a single Norfolk Broad, which was quite widespread in the early Flandrian. This could be more a comment on habitat than on climate. No one has suggested that the Elm Decline mainly affected, say, *Ulmus laevis*, hardy at Kew but confined in its present-day range to central Europe.

However, the proof of the variation in flora during the early post-glacial period is adequately given in the very technical form of phytogeographic analysis, as Godwin demonstrates.

Beech and hornbeam, which entered, or re-entered, the country late in the Atlantic Period, at about the same time as the first forest clearances, did not, it appears, enjoy a wider range than they do today – if indeed their distribution was ever entirely natural. A people capable of thatching huts, decorating pottery and tending animals was obviously capable of planting two very useful trees.

I have not discussed theory *g*, that modification of the soil made it unsuitable for the survival of elm woods. Only a really large incursion of the sea, as postulated by Velikovsky in his survey of possible catastrophes, *Earth in Upheaval* (1955), could have altered the soil rapidly and selectively enough to have caused the Elm Decline. It is still possible, but it belongs under *a*, the Flood theory. An alder and willow decline, and that of many other plants, would have accompanied such an occurrence. The existence of coastal submerged forests all around Britain proves nothing except that there was once a lot of trees, which we know. I suppose a theory *h* would postulate a few decades of very strong gales, knocking down the shallow-rooted elms.

Winifred Pennington (1969) discusses the formation of *Sphagnum* peat and the natural process of debasement of the soil by leaching during an interglacial. This whole interesting question I must admit is beyond my powers to even outline. One factor must be that 3000 years of high forest in a wet climate must surely produce considerable amounts of humus, with perhaps excessive acidity and water retention. But high forest of whatever nature is always said to be self-perpetuating if undisturbed. Some woodland must have been lost to adjacent mires, especially when temperatures began to go down.

As Pennington remarks: 'In the second part of our present interglacial . . . man has for the first time become a dominant factor in the ecosystem. . . .' The Elm Decline is closely linked with the settlement of modern man in this country. Other forces must have contributed, much or little. But we know as little for certain of the natural conditions around 3000 BC as we do of the reasons which brought settlers here at this time. Over-population of the Danube forest? It seems unlikely. Perhaps some of the people came in lovely ships from Phoenicia, early pilgrims from Jehovah knows what cultural oppression, or even deported dissidents. I like to think I should have been one of these, exploring the British forest.

The Historical Period

Elms coppiced for useful timber or planned into great schemes of landscape: both are part of our history

The Flandrian pollen zone VII, which is split into two important halves by the Elm Decline, ends at about 500 BC with the climate settling into the familiar cool, wet Oceanic one of our own time. Zone VIII begins as the Bronze Age moves into the Iron Age. Zone IX witnesses the gradual reduction of all trees as farming and iron smelting progress and population rises – very gradually.

Zone X, which is being laid down at this moment, will probably show a considerable increase of spruce, pine and other coniferous pollen including many North American species. Near the lower boundary of the zone will be detected an unbelievably sudden drop in elm pollen. Fortunately for future paleobotanists there will be, no doubt, adequate printed explanation.

When the Romans came to Britain they found in the south-east a system of arable farming, while the Britons in the north and west, lived, they said only on meat and milk. The Romans brought vines, mulberries, walnuts, figs, plums, cherries medlars, and peas, radishes, celery, fennel, coriander and dill. They planted firs and stone pines grew cannabis and opium poppies, and cultivated woad, flax, belladonna, henbane and vervain. The last three were possibly for rheumatism, sleeplessness and to induce hallucinations, in that order – but vervain was also an altar plant. The Romans may actually have brought the

The church of Elm, near Wisbech, dates from the thirteenth century. Perhaps it should now become the memorial church of the elms

which Julius Caesar noted was used by the Britons to dye their skins blue.

Considering the practical purposes of most of their importations (even stone pines provide an edible 'nut', and firs make all-season wind-breaks) it is hard to see why they should bring elms of any kind to a heavily wooded country. But some writers still believe that all our field elms were introductions either just before or during the Roman occupation. If the Romans insisted on particular species of elm to prop their vines, and *Ulmus glabra* really was the only native, that would be another story – one that most botanists do not believe.

The Romans are supposed to have used elm for the hinges of doors, the elm post turning in sockets of stone. But that seems to be all we know. When they left, whatever was the state of their British province, it has become clear that they had made it into a great supplier of grain. There would be some place for elms in the arable fields, and some use for the timber, whether coppiced or free-standing. The most continuous demand in the Roman settlements was for fuel, for heating, grain drying and industry – iron-foundries, brick kilns and potteries. Elm probably held its place in the Roman coppices to maintain local supplies of fuel over several centuries.

The Anglo-Saxons, who charted their land by perambulations of boundaries, left some indication of the tree flora. The tree often mentioned as a landmark is a thorn. The tendency of hawthorns to be left isolated in the open is well known – they are either magical or the cattle will not eat them, or both. Hawthorn hedges, also, were already popular. But in the trees mentioned in 150 charters listed by Rackham (1976) there are no elms at all. These charters were for south England and particularly Worcestershire, where perhaps the elm ('the Worcester Weed') was so common a sight that no one could possibly use it for a landmark. The Anglo-Saxons were builders in wood, the inventors of the frame house, and if they did not find a use for elm it would be surprising.

One charter at Colchester mentions cleft elm trees – they were only notable because they were curiously split. There are one or two records of fines imposed for cutting elms.

Place names for elm are a most unreliable guide: there are Eller and Ellen for alder, and names like Elli, Ella and Aethelmaer shortened to Ell. Perhaps we can accept Elmleys and Emleys (Worcestershire, north Kent and Yorkshire) to be elm woods, and Elmsett (Suffolk), Elmstead (Essex) and Elmsted (Kent) to be homesteads specially marked by elms. A northern name for the elm was holm, presumably with the 'h' dropped: could it come from the Norse *helm* for a rudder? 'Holme' further south is an island or a lake – it is all rather vague.

Elm is itself Old English, an adaptation of *Ulmus*. *Wice*, or corruptly Witch Elme – as distinct from Witch Hasell – was the name of the smooth-leaved elm in Essex, where real wych elms are few. (For the Celtic names of wych elm, see p. 59.) The fact that there is no Celtic name for field elms is used as a weak argument for their being of Roman introduction. But even Germany uses *Ulme* (as well as *Rüster*).

As for the Normans, we know nothing of their elms apart from places called Elmdon, Elmswell and Witcham. *The Domesday Book* tells us nothing, except that villages had woods and patches of 'underwood' – unenclosed, but somehow measurable in the Midlands, and assessed on the size of the pig herd in the south-east. The swine statistics may already have been out of date when they got into *Domesday*; but they strongly suggest oakwoods and beechwoods.

A map of places *not* mentioned in *Domesday* would seem to indicate that the virgin forest had almost disappeared by this time, at least in lowland England. While woods are rarely mentioned by name, and trees never, *Domesday* pictures a countryside settled thickly by communities whose whole economy was linked to trees and woods: those whose woods were already gone were forced to depend on coppices. We may guess that every tree was used for something, and that picturesque groups of elms were few indeed.

Medieval coppices

Whether or not a village had any woodland, or wood pasture, and especially if it did not, coppices were essential to supply the constant demand for moderately straight pieces of wood. The great areas of native hazel that showed in the pollen charts of the Mesolithic Age were everywhere turned into a sort of living timber yard, where wood of whatever size required was available when it was wanted. Besides hazel, the most common, nearly every other broad-leaved tree could be included in the coppices, including oak.

A coppice tree is one reduced to a many-stemmed, tall, bush-like growth by repeated cutting at ground level. It never forms a trunk or bole, but the shoots grow vigorously over and over again from a steadily enlarging base or stool. The stool frequently builds up into a woody mound, but it can be under the surface. Coppices in hilly, exposed places have been mistaken for scrub in the past, and the climate blamed for the low stature of the trees.

Coppices can be cropped in strict rotation, say every seven, ten or even fifteen or twenty years, but the medieval carpenter seems simply to have selected what he wanted and worked it green, so that a stool might eventually bear shoots of various ages and lengths. A coppice can be re-converted to a stand of timber trees by allowing only one stem to grow. This explains the slender trunks of hybrid elm in the picture of part of Hatfield Forest.

A law of Henry VIII in 1543 discouraged the woodman from cutting ash and oak in the coppices. These trees were to be allowed to grow up as standards, stores or princes. The custom was probably well established by the time the law was passed. The pattern of coppice-with-standards can be seen frequently today, much neglected. An earlier Act, of 1483, allowed landlords to fence a part of a coppice temporarily for seven years after cutting, to keep out deer and the commoners' cattle and pigs, all of which eat young tree shoots. Again, the practice probably existed before the law, which merely put the matter out of dispute. Copse-banks, sometimes very ancient, are the remains of such intermittent fencing.

Pure elm coppices are very rare. Rackham (1976) describes the oldest part of Groton Wood, now the property of the Suffolk Trust for Nature Conservation. It is 'dominated by huge stools of small-leaved lime, accompanied in wet hollows by elm (not an invasive elm, but one of the coppicing sorts that sometimes go with lime in Suffolk and may have done so since the Elm Decline)'.

The decline of lime, however, has been more complete than that of elm. Before elms began to invade woodland from the hedges, they may have escaped from such coppices into hedges and woodland, progressing by suckering, while the seedlings of lime were eaten by mice, birds, rabbits and intelligent children.

U. procera is not a coppice tree – though indeed it makes a sort of coppice of its own, usually seen confined to the hedge on either side of the tree. The hybrid which formed part of a large coppice at Hatfield appears to be a type of Dutch elm, though botanists would probably not allow this name. *U. glabra* itself is an uneasy tree in the coppice, requiring a great deal of space for stems which shoot upwards, then sweep down, as with the mature tree. Two large wych elm stools remain at the edge of a cant of hazel in Hatfield Forest.

Pollards

The habit of combining pasture with woodland, which may go back to the late Stone Age, produced a stabilized pattern in medieval times. The

pattern was very variable according to the amount of woodland remaining and the number of beasts which cropped the grass. Seedlings, shoots and young trees also got eaten, and the tendency of all land subject to common rights was to become treeless, unless grazing dropped to less than about one cow to 4 acres (1.5 ha). Commoners' rights to firebote, housebote, hay (hedge) bote, ploughbote and estovers (dead wood) did not in theory reduce the woodland if growth and regeneration continued. Turbary – the right to cut turf for fuel – probably did more harm in some areas.

To ensure the supply of small-section poles, trees were lopped or polled, not cut down. Pollards are a sort of aerial coppice, above the reach of animals' jaws. Lines of such trees of almost any species are a mark of old common land; sometimes of boundaries; and of well-kept villages of any age. A lot of modern pollarding in towns is of course just misplaced tidiness. It is easy to take a modern crosscut saw up a ladder, rather less so to poll a tree with the axe, slasher or brushing-hook. (But saws have been in use continuously since the Iron Age and before.)

Many very ancient trees are pollards. Pollarding, like coppicing, stimulates rapid growth of new wood, but greatly slows down the growth of the trunk. Added to this, most really old trees could not have survived while carrying their full load of branches – some have been naturally pollarded in storms.

Pollarded trees on farmland were once much more common than 'maidens' (untouched trees). Eighteenth-century writers complained of the too frequent ugly boles of ashes – this is now a rare sight. Recently pollarded elms can still be seen in the Stour valley in Suffolk and in parts of Essex.

Top **Landscape with lopped elms (U. procera), Oxfordshire**
Above **Roughly managed wood of Plot elms, Cambridgeshire**
Opposite **Massive ancient stool of wych elm in Wiltshire**

Even in such wooded parts of the country, farmland trees were five times as thick on the ground in the early eighteenth century as they are now, according to Rackham. He describes also (1976, p. 130) the partly deserted village of East Hatley in Cambridgeshire, where the lumpish, ancient boles of pollarded elms mark the lines of old paths and plots. The trees belong to as many as ten different clones distributed in a pattern which appears to relate them meaningfully to the archaeology of the place: ponds, pits and ditches full of nettles.

Elms cut in this way are usually hybrids, and show their wych elm parentage in the large leaves of the still juvenile shoots. Even though elm bark-beetles are not supposed to lay their eggs in shoots of less than ten years of age, some of the pollarded elms in Suffolk are dead.

Shreding

An alternative way of lopping a tall elm was by shreding, shrieding (shrieving?) the side shoots, leaving the crown intact (see p. 65). This treat-

ment, especially for *U. procera*, avoids rotting of the trunk caused by exposed cuts, and was probably most used for reaping the foliage for fodder. It is now very rare to see an elm, or any other tree, shreded in this country.

Grown close together in a wood, elms of normally branching or spreading habit, like any other tree, grow as tall straight trunks with full crowns. A few trees left standing at the edge of a clear-felled wood will show this effect – often to be seen where a motorway has bisected a wood.

The domestic elm

We must conclude, without any real evidence,* that elms were planted, or left to grow, around the earliest permanent habitations, at least in south-eastern and south-western England. That is their habit and their reputation. The elm is a difficult tree to eradicate, and in any period of woodland economy, a difficult one to do without. Where it now decorates a village or homestead it must, even 1000 years ago, have supported the numerous functions for which a whole range of specialized metals are now used: wheels, solid or spoked; troughs, water containers and pipes; dairy equipment; mangers and bars for cattle – in fact any timbers that had to avoid splitting, take hard knocks, or remain sound under water. The foliage was fodder for the beasts in a dry summer, and perhaps was eaten by the people in early spring. The crown provided shade, and, remaining green until late in the year, was some protection from storms.

Rackham, who seems as much guided by a highly sensitized intuition as by undeniably thorough research, has christened the elm the main tree of settlements. If you come across a group of elms in a wood, he says, look for signs of an old settlement. The poets have told us that elms are witnesses of our lives and deaths.

If late Stone Age farmers chose elm groves for their clearings, then left the sites when the soil seemed to lose its fertility, it is more than likely that other men, less concerned to grow corn, took over the clearings after them, later coppicing the new growth of hazel or sallow and at least tolerating the elms which we know would grow up again. The clearings might have been abandoned again and again over five or ten centuries, but being the natural places for habitations, above the flood line of the rivers – and the natural bases for further clearing of the forest – they remained, elmy places that eventually got onto the map.

From all such settlements, wherever elms grew, they would have spread slowly, as they do now, into neglected corners, ditch banks and hedges. Their spread was assisted by planting, not perhaps very widely in medieval times; but, increasingly, as carpentry became more sophisticated, water mills were established, coffins were demanded, pumps and drains were built, and rivers navigated.

So much we may speculate, without, I hope, being accused of romancing.

* But see Richens, R. H., 1967, 'Essex elms', in *Forestry*, no. 40, and his earlier articles listed there.

Enclosures

The more land was enclosed, the more hedges were made and the modern habitat of elms increased. There are now about half a million miles of hedge in Britain, in spite of a tremendous move towards larger fields to accommodate the post-war generation of farm machinery, to increase acreage and to reduce maintenance. A government subsidy to hedge-removers ceased in 1972.

The Cistercian sheep farms were the first important enclosures. In the lowlands the process began in the fifteenth century, and was complete by the middle of the nineteenth century. The old pattern of open field farming can be seen at Laxton, Northamptonshire. The great sheep farms of the moors and mountainsides, once enclosed by those heroic miles of dry-stone walls, are now enclosed only in reverse by the lowland fields and the edges of Forestry Commission conifer belts. Startlingly luxurious patches of woodland grow in the moors, wherever a bit of land is accidentally protected from sheep, deer, hares and rabbits. Wych elm may be found in such oases up to about 1000 feet (300 m), frequently near habitations, not so frequently in wild, steep valleys which tend to be given over to birch, rowan, hazel and alder.

On the position of the elm in southern England, John Aubrey, writing about Wiltshire in the late seventeenth century, says:

'I never did see an elme that grew spontaneously in a wood . . . Anno 1687 I travelled from London as far as the Bishoprick of Durham. From Stamford to the Bishoprick I saw not one elme on the roade, whereas from London to Stamford they are in every hedge almost. In Yorkshire is plenty of trees, which they call elmes; but they are wich-hazells, as we call them in Wilts (in some counties wych elmes) . . . A great many towns are called Ashton, Willoughby, &c. but not above three or four Elme-tons.

'In the common at Urshfont was a mighty elme, which was blown down by the great wind when Ol. Cromwell died. I saw it as it lay along, and I could but just looke over it.

'Since the writing this of elmes, Edmund Wyld Esq of Bedfordshire assures me that in several woods, e.g. about Wotton &c. elmes doe grow naturally, as ashes, beeches &c; but quaere, what kind of elme is it. . . .

'Wich-hazel in the hundred of Malmesbury and thereabout, spontaneous. There are two vast wich-hazel trees in Okesey Park, not much lesse than one of the best oakes there. . . . When I was a boy the bowyers did use them to make bowes, and they are next best to yew.'

But Aubrey was not a great lover of the elm country:

'In North Wiltshire, and like the vale of Gloucestershire (a dirty clayey country) the Indigenae or Aborigines, speake drawling; they are phlegmatique, skins pale and livid, slow and dull, heavy of spirit: hereabout is little tillage or hard labour, they only milk the cowes and make cheese; they feed chiefly on milke meates, which cooles their braines too much – and hurts their inventions. . . . In Malmesbury Hundred, &c (ye wett clayy parts) there have ever been reputed witches.'

These were the old enclosed lands of Wiltshire.

The general intention of landowners in seeking to enclose their land was to make it profitable: while many trees were cleared in the process many were also planted, and a good proportion of these were elms. The hedges, usually of hawthorn, were laid periodically and in them timber trees, most often ash on high ground and elm on lower ground, were allowed to grow up. New hedges in the eighteenth and nineteenth centuries often had elms planted in them at regular intervals; their offspring are somewhat more irregular and often in close groups of two or three.

Already, in the middle of the seventeenth century, John Evelyn raged at the landowners not only for cutting but for utterly extirpating 'all those many goodly woods and forests, which our more prudent ancestors left standing for the ornament and service of their country'. The devastation had become, he said, epidemical, and the Navy would soon be short of essential building material. He invoked the fear of invasion to persuade people to plant trees. His work became famous, and we can be sure that his advice was taken.

All sorts of trees had to be planted, for the destruction was widespread. Evelyn gave most of his attention, in his *Sylva* (1664), to those that were most depleted and those that might be most useful. The chief of these was the oak, as everyone knows, but the order in which Evelyn placed his trees is instructive.

Dry	*Lesser species*	*Aquatical*
Oak	Service, Maple,	Poplars and Aspen
Elm	Lime, Hornbeam	Alder
Beech	Hawthorn	Willows including
Ash	Birch, Hazel	Osiers and Sallows
Chestnut		
Walnut		

He would also encourage fruit trees – he was in favour of planting them between the oaks in the Royal Forests; deer and cattle might feed, too, he thought. Or oaks might be planted in the fields, about 100 feet (30 m) from the hedges. You may often see them so on estate farms.

Beeches would grow in valleys and on stony, barren soil, and on the sides and tops of hills, especially chalky ones. Ashes, if not lopped and knotty, might be valuable, but their roots would be in the way in farmland and their shade was thought to be injurious – perhaps another way of saying that the ash is greedy with soil. In hedges they would thrive exceedingly, and in 'plumps' 9 or 10 feet apart.

Chestnut he would also put in the hedge, but this is rarely seen: more often now they are in small stands on well-run estates. Evelyn praised the product of the chestnut coppices. Walnuts, planted at 40 or 50 foot (12 or 15 m) intervals would make graceful avenues to country dwellings; this plan may have been widely followed, but the acute demand for walnut, in the century and a half after, probably saw the end of most of these avenues, however graceful. A very severe frost in 1709 killed off a great number of walnuts. The wood was considered essential for gun stocks, and during the Napoleonic Wars as much as £1000 was paid in England for one large tree (that would be at least £10 000 today).

About the elm, second on his list, Evelyn's advice was complex but clear.

'. . . the elm delights in a sound, sweet and fertile land, something more inclined to loamy moisture, and where good pasture is produced; though it will also prosper in the gravelly, provided that there be a competent depth of mould and it be refreshed with springs. The elm does not thrive in too sandy or hot ground, no more will it abide the cold and spongy; but loves places that are competently fertile, or a little elevated from these annoyances, as we see in the mounds and casting up of ditches.'

OF FOREST-TREES. 113

CHAP. IV.

The ELM.*

1. ULMUS, the ELM. Of this there are four or five sorts, and, from the difference of the soil and air, divers spurious. Two of these kinds are most worth your culture, viz. the Vulgar, or Mountain Elm, which is taken to be the Oriptelea of Theophrastus, being of a less jagged and smaller leaf; and the Vernacula, or French Elm, whose leaves are thicker and more florid, glabrous, and smooth, delighting in the lower

John Evelyn

The ulmarium

Despite its 'aspiring and tapering growth' the elm's shade would not harm corn or pasture: but the root suckers should be cut out. Because of this habit of suckering, holes could be chopped in the main roots of a vigorous elm, the wound propped open with stone – and then by covering them with earth again one tree could produce a fair nursery of shoots. After two or three years the young trees could be separated from their parent and planted in the *ulmarium*, or in clumps, 10 or 12 feet apart, or, even better, in hedgerows: 'For the elm is a tree of consort, sociable; and so affecting to grow in company that the very best I have seen do almost touch one another'. An elm, Evelyn says, does not thrive in the forest where there is no room for its roots to spread out, as there is in hedges and avenues. There was no easier tree to transplant, even after some years' growth.

Have your *ulmarium* if you liked the idea, but above all plant elms in the hedgerows of your best land. Useful timber would be assured; lopping the side branches would provide firewood; and the leaves would feed the cattle in winter and in scorching summers when fodder was dear. This was the model for the Enclosure Acts hedges.

Batty Langley's twenty-year plan

It was one thing to persuade the king (Charles II) to restock his forests with oaks for the maintenance of our wooden walls, and quite another to make landowners see the value of planting, when they might wait thirty or forty years for any profit, or be in their graves. Oaks, at twenty or thirty to the acre, used a lot of land, even if coppice trees were grown between them. There was a way of making a good profit in twenty years by planting elms, and this was set out, with elaborate year-by-year accounts, by Batty Langley, a landscape gardener, in his *Sure and Easy Method of Improving Estates*, published fifty years after Evelyn's *Sylva*. Large, free-standing oak timber was in demand, not close-grown and straight, but full of the 'crucks and knees' so desired by the shipbuilders. But there was a good market for elm poles.

Your oaks might be worth 10s. a tree after twenty years, but elms would be worth more than 20s. each, and three times that after forty years (when an oak is still young). Elms could be planted 108 to the acre. For the first ten years or so 'good crops' could be grown between them (their side shoots being cut). After ten years the land could be put to pasture as well.

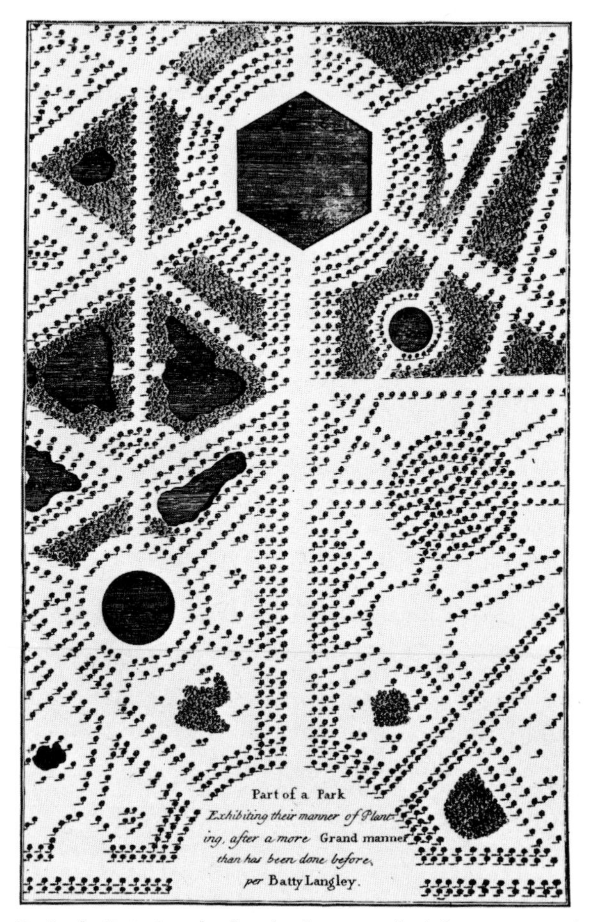

Design by Batty Langley for a landscape garden 'after a more grand manner than has been done before'

Renting out your land for twenty years might make £11 per acre if all the proceeds were invested. Planting oaks would make £13. But elms would produce £84 per acre, over and above the crops grown between them, and the fodder, and the firewood. The rent of the land is estimated at 5s. per acre, and Langley takes into account the interest on all his expenses including the initial 'trenching of twenty Acres of Land two Spit and a Crumb at 6d. per Rod . . . £80'. He does not include the price of the plants. These would be less than 20s. per hundred, but he recommends instead taking suckers out of the woods and hedges (1s. 6d. per packhorse load is quoted elsewhere) and keeping them on nursery beds for two years. The final expense, felling, would be covered by the sale of the tops.

Langley also recommends planting English elm in hedges. Saplings for the hedge must have their leading buds 'displaced'. The young plants can then be 'thickened at their bottoms' by pinching off the extreme buds of their lateral shoots. Elms were to be planted out in October or November.

These powerful recommendations of 1664 and 1741 do much to explain how elms came to be, in many southern counties, the most important single tree in the countryside. Even so, in 1968, Darrah and others (in 'Woodland elms in Wessex,' *Forestry*, no. 41) expressed concern that the demand for elm timber in Wessex might exceed the supply. They noted that when surveyed in 1955 there were about 800 million cubic feet of timber in hedges, paths and small woods, making twenty-one per cent of the total standing timber; elms were twenty-one per cent of this.

The authors expressed their regret that so much finely grained and useful wood should end up out of sight in the graveyard. Now, of course, most of the Wessex elms are diseased, and only a fraction of their timber will be recovered. Perhaps one outcome will be a new appreciation of the beauty of elmwood, ousting the present fashion for pine.

English elms at Aranjuez

Avenues of elms

Famous avenues were planted by Philip II of Spain at Aranjuez: they were of elms from England, perhaps intended to make Mary Tudor feel at home. The trees grew very tall, being only 15 feet apart; the side branches were stripped to the tops. Double rows of trees, 6 feet apart alternately, straddled irrigation channels, flanking walks and rides about the palace.

In spite of this elegant regimentation, and in spite of their reputation for barrenness, the elms, which were *Ulmus procera*, produced seed which was taken all over Europe, it was said. Spain is not noted for elms of any other sort.

When in the 1660s Evelyn blamed the spread of tillage for the dearth of good trees, he tactfully failed to mention that nearly 1000 deer parks in England and Wales had done a good deal to inhibit the spontaneous regrowth of the natural forest. Fallow deer had been kept by the gentry since introduced by the Normans. The object of a medieval deer park was to ensure a supply of fresh meat, with the manly exercise of killing it in your own bit of forest. The fences to keep the deer in and the serfs out were of 12 foot oak stakes, which accounted for quite a lot of oaks; deer, even at less than one per acre, eat every leaf that they can reach. After a century or two, a deer park would consist mainly of very old oaks, dead elms and small hawthorn and holly bushes. Many owners fenced off bits of woodland, instigating the tradition pattern of English parkland, but some large parks were described in the seventeenth century as deserts, with little timber, though on good land. Many owners had spent their time chasing Roundheads instead of deer, and the herds had no doubt increased.

Perhaps because of Evelyn, and certainly partly because King Charles II had got a colleague of Le Nôtre to lay out St James's, and, perhaps a little because the ladies wanted it, a fashion began before 1700 for ornamental tree planting.

Many alien tree species nowadays familiar in parks were tried out, and Scots pine returned to England, with European larch, fir and spruce. But oak, elm and lime were the basic pigments of the landscapers' palettes. Oaks were planted mainly to replace the losses in grander chases, purlieus and parks, but elms and limes were for the avenue. They were often imported hybrid elms, and usually the limes were the now ubiquitous *Tilia × europeae* (*T. vulgaris*). They were planted for their delicate youthful tracery and spray, and would probably have been hated as crowded, heavy old trees, turning ladies' complexions green, and accumulating dank airs and insects.

The method of planting somewhat reflected the formal Dutch gardens of the period, but on a gigantic scale. The intricate radiating and interlocking geometrical patterns of avenues which were planted out over many square miles must have deeply confused everyone on the ground. They also provided plenty of solid work for two famous illustrators of gentlemen's country seats, J. Kip and L. Knyff. Topographical writers of the time were impressed with the transformation, the beauty – and the expense. Perhaps the designs expressed in some way the new enthusiasm of the leisured class for 'experimental philosophy', and they were no doubt partly a reaction to the disordered landscape of the 'ancient' countryside, where the only pattern was the interminable fragmentation of woods and fields, punctuated by windmills.

Grandiose as these schemes were, the most

An early example of suburban street planting – elms at Bonchurch

ambitious pioneer of English avenues must have been the Duke of Montagu, who returned from being Ambassador at the Court of Versailles to plan a seventy mile avenue from Boughton in Nottinghamshire towards London. Some miles of this early precursor of M1 were completed, and many more avenues of lime and elm on his estate are said to have totalled the same mileage.

By the time all these plantings had begun even to approach the maturity we now expect of an avenue of trees, the fashion had changed; formality was out, and landscaping had come in.

However, avenues were popular, and the landscape gardeners allowed the gentry to keep them, provided they were over undulating country which would break up the tiresome geometry.

Though the work of the landscape gardeners was noted for its informality at the time, it was in fact fairly rigid when first practised. It is now overgrown by the standards of its original planners, and all the more impressive for that.

Blenheim, Cassiobury, Hampton Court, Kensington Palace, Hyde Park, Richmond, and Windsor were all improved by Bridgeman, who died in 1738. He also worked on Houghton Hall, Norfolk; Stowe, Buckinghamshire, and Wimpole Hall in Cambridgeshire (near Royston). William Kent followed him at most of these places – Chiswick and Rousham were his work too – and Capability Brown added further and more far-reaching improvements to most of them and about 150 others, including all the famous parks. Repton came in at the climatic stage, often content merely

to remove a few trees and build the odd village.

Avenues are only parts of the great schemes of English parklands which have one thing in common: the ambitious love of their incumbents, backed by enormous wealth. They do not look 'designed', and that lovely balance of land and trees owes as much to our benign climate and the sustained good taste of their owners as to any plan.

Stowe's mighty avenues of elm or lime lurching over the little hills of North Buckinghamshire are greater than they were designed to be, but they are at the end of their life now; the three straight miles of elms planted in 1700 at Wimpole are now felled, to be replaced shortly by limes.

Blenheim, planned in bold masses round its artificial lake, also has its avenue – eight rows of elms for a mile westwards from the Column – added after the time of the improvers, in 1887. Elm disease has touched the edges only, so far, of this fascinating mechanical forest of highly consistent hybrids.

Five thousand years since the Elm Decline, we see that the elm population is as mixed and intermixed as the human population of these islands, and the real study of elms begins, perhaps, where botany and taxonomy fail, with the study of the people, cottagers or dukes, who have made the landscape their own and planted, lived with, used or admired the elms.

Elms are the predominant tree in this park at Brighton.
Below **Blenheim Park; a great double avenue planted in 1887**

Part 4
The Timber Tree in the Hedgerow

The Village Carpenter

There was one good reason for growing elm trees close at hand: the technical but indisputable one that most types of the wood cannot be cleft. Wedge and sledge, axe and beetle, froe and mallet, it defeats them all. The felling axe just bounces off the end grain of most logs of English elm. Wych elm and some hybrids, however, can be cleft. Medieval carpenters cut, trimmed, shaped and, where necessary, cleft their wood in the coppice, and most wood craftsmen up to modern times have reduced their material to manageable proportions before bringing it out of the woods. Many, such as hurdle makers and chair bodgers,* clog sole cutters† and less specialized makers of stakes, rails and poles, spent their whole summers in the woods. But elm was usually needed in large pieces – squared, it is true, with the axe, by a skill that died out in the mid-nineteenth century, but needing to be sawn.

Oaken clapboards, panels, rails, struts of up to 10 feet, gate parts, down to laths and slats, and especially wheel spokes, ladder rungs and barrel staves, were all made better and stronger from cleft wood than from sawn.

But English elm must be sawn, for its grain is twisted and crossed: that is its particular value, that and its durability under water. Its particular uses are for work that must not split. Nothing else will do, for instance, for the hubs of wooden wheels: hollowed to take the axle pin and with at least twelve holes all round for the spokes to be driven in, they have to withstand the sudden grip of the cooling iron tyre, then the weight of cart and load, and all the shocks and bumps from the road.

But elm is a hard and weary wood to work, and oak is preferred if it can do the job at all.

At some stage, and at different times in different places, the carpenter came to rely on planks of timber for most of his work rather than on sticks and hewn or cleft wood. For two centuries of the finest craftsmanship the planks were produced by two men working one above the other in a sawpit.

*Chair bodgers made the sticks for stick-backed chairs in the beechwoods above High Wycombe.

† Clog blanks were produced in the Welsh alder woods to supply the makers in Lancashire.

The logs or butts of oak and elm lay in the carpenter's yard until they were needed: oak might lie for years and only improve, elm must be used after a year or it would be weakened. A 'white grain' effect creeps through it, which only an expert can see. The butt was squared on one side with the axe and marked for the saw cuts by the accurate flicking of a stretched rope soaked in lampblack. It was then manoeuvered, by levers

and iron ring-dogs, until it was face down on a series of timbers over the pit. A pit roll (see drawing), hexagonal in section, was used with a long lever to move the butt forward or back over the supporting timbers. While being sawn, the butt was secured by irons which had right-angle spikes at each end, driven into the pit timbers.

The lines for the top sawyer were projected from the squared side (now facing down into the pit) onto the top of the log. The top sawyer stood on the work, the bottom sawyer in the damp, brick-sided pit, with sawdust constantly showering him and the risk of a ton or so of wood falling on him. The saw was sharpened with a half-round file, tooth by tooth, by the top sawyer, while the bottom man prepared the next butt.

Several parallel cuts were made to produce planks from 3 inches down to ½ inch, and even veneers. The bottom handle of the saw was detachable, so the blade could be removed easily.

The planks were shooked (packed in sets) against poles with crossbars to dry out, and were then stacked horizontally, separated by thin pieces of wood. At this point elm boards begin to warp – they were sometimes weighted down to prevent this.

Sawing a tree into coffin boards (1 inch and ¾ inch thick) would take two men about one week.

Planks and roughly shaped blanks were supplied to furniture makers, and blanks made for wheelwrights, for hubs and felloes (rims). Elm was used extensively in routine repairs to water mills. The village carpenter himself made and repaired many objects of elm wood, as below.

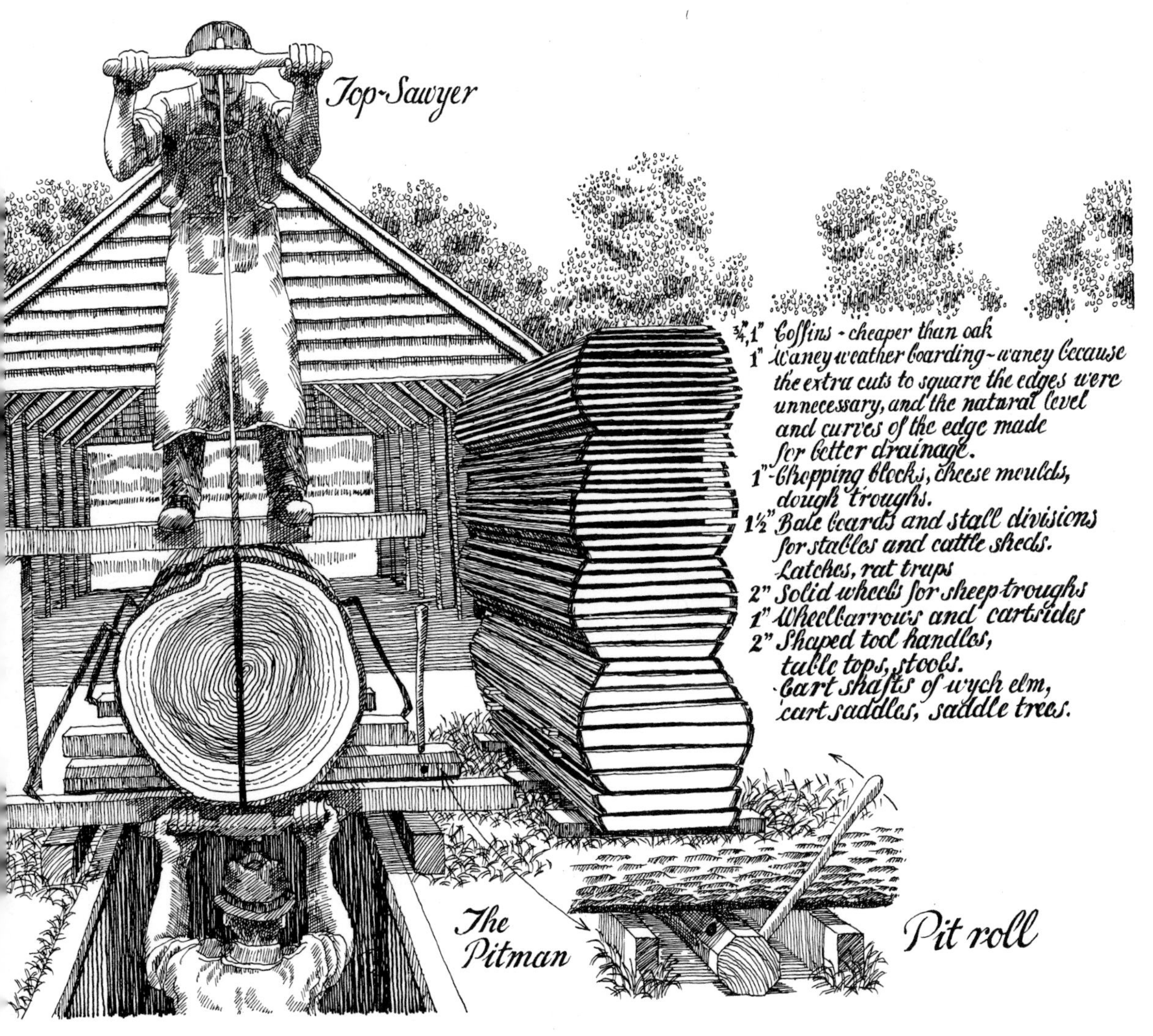

Waney-edged Weatherboarding
LAGE
PENTER
Sawn logs
Coffins
Sides to carts
Wheel hubs
Curved grain Felloe
Shafts-Wych-elm
Wedges
Bowls
Ammunition boxes
3/4"
1/4"
Chair seats
Water pipes

Feather-edged
Weatherboarding
Water-
wheel
Stable divisions
Three-yard
tip wagon
Cattle trough
Bench
2"
Bakers
dough trough
Wheel barrow
Washing dolly
Ball-feet on furniture
Chopping
Blocks
Cheese
moulds
Plates
Bellows
Egg cups

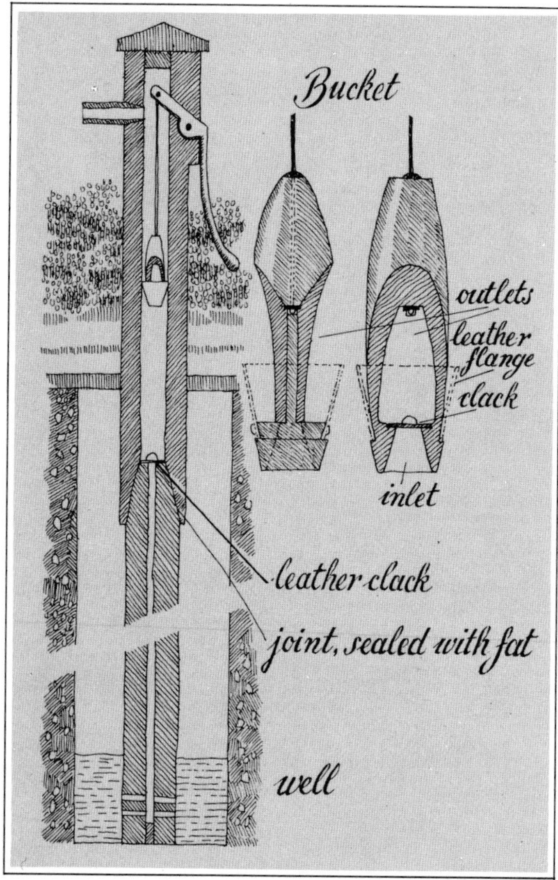

The bucket pump

The extraordinary skills employed – and the triumphal success of elm wood – in the building and setting up of wooden bucket pumps are described from first-hand experience by Walter Rose of Haddenham, Buckinghamshire, in his book *The Village Carpenter* (1937). A bald summary cannot convey the skill and accuracy of handwork required to raise a plentiful supply of water up to 20 feet from the well. In theory 34 feet is the maximum height for a suction pump, but 20 feet is good in practice by any standard. In Rose's time these wooden parish and farm pumps were being replaced by metal pumps, but these had lead pipes which burst in the winter.

The entire pump was constructed of elm, three pieces of the best leather, a few nails, two lead weights, and a heavy handle of iron, with a pin and rod. The body of the pump and the longer pipe which reached the bottom of the well were squared baulks, bored in the carpenter's yard and fitted together in the well by a taper-and-hollow joint. The lower pipe had a 2 inch bore, the bottom end being plugged and water admitted through small holes in the side, thus excluding frogs and detritus. The upper part had a 5 inch bore to take the moving bucket.

The bucket could have been turned from no other wood but elm, for it had to be a mere shell, and stay wet in the pipe for years. A leather flange or sleeve was used to make it airtight as it rose in the pipe. A non-return valve, called a clack, of leather weighted with lead, was fixed at the joint between the pipes to hold the water in the pump. Another valve in the bucket opened when this was lowered, admitting water through a hole in the bottom. When the bucket was raised to the head of the pump, water escaped through large holes in its sides. An almost regular flow came from the outlet as water was also displaced by the bucket falling at the next stroke.

For making the pump, growing trees were chosen with great care; they were cut and worked at once, for no seasoning was given to elm which had to stay wet.

Boring the pipes took two men about a week, using a long auger with a spoon bit, or a shell auger, supported on a carefully lined-up stand. A screw-pointed auger would wander along the grain and could not produce a straight hole. The long pipe was lowered down the well, and a man descended to fit the valve, then to guide the upper pipe into its socket as it was lowered, and sealed it with a bowl of hot sheep fat. The pump was then primed with water from the top, which had a removable lid (of elm wood). Rose describes how the farm people would gather to see the new pump tested. An intermittent flow would mean that

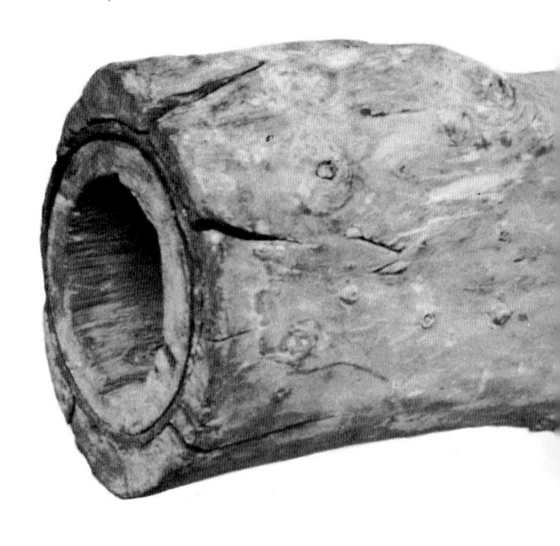

An Elizabethan woodcut (Agricola, 1556) shows an ancient, possibly Saxon device for raising water. Below **An elm branch which served as a water main, with a 5 inch bore, for over 100 years in Millwall, and a stop valve dating from 1698**

there was a fault somewhere: a trickle after the handle was released meant success. This was the 'climax to many days of hard labour, skill and care'. Its makers saw in the pump 'their humble contribution to the welfare of mankind'. And we see why the hedger was careful to let the elm grow up.

The pipes for the wooden pump employed a method used for miles of water pipes under London. Wooden pipes laid in 1613 for the New River Scheme were dug up in 1930 still sound. Such wooden pipes can often be seen in museums. So many elm pipes (and perhaps oak, alder and other woods) were used in London that the process was mechanized in the eighteenth century, water wheels driving the auger and racking the log forward. Edlin, describing the making of water pipes in his *Woodland Crafts* (1949) adds a philosophical paragraph which might apply to many aspects of our modern economy, where each community depends for all its needs on an ant-like scurrying of carriers, all using expensive energy.

'If the boring of water pipes by hand strikes us as a laborious and uneconomic process, our present practice of digging up coal and iron ore, transporting them over great distances to blast furnaces and ironworks, and then carrying the finished pipes to their final resting places, would have appeared equally absurd to the old carpenters, who could produce pipes from trees growing on the spot.'

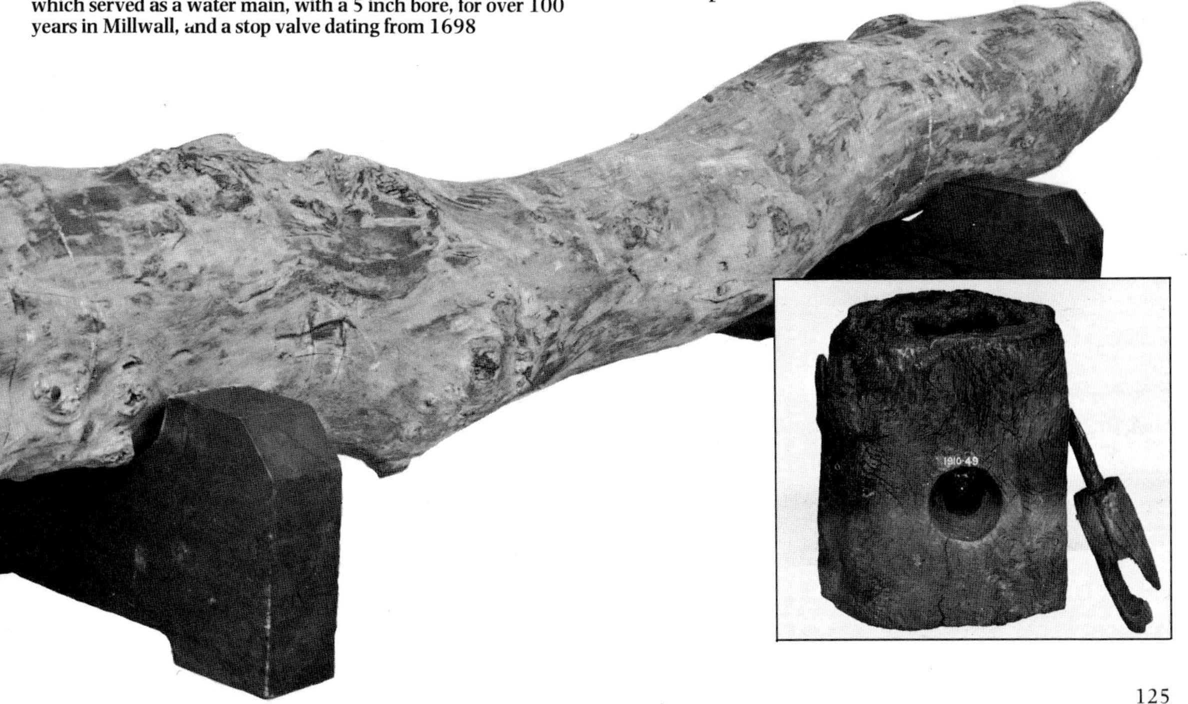

The Elm in the Timber Yard

While the carpenters of a thousand villages in the Victorian era followed methods of craftsmanship slowly evolved over several centuries, local industries became centralized – particularly furniture-making at High Wycombe. Here beech forests supplied the vernacular tradition of stick-back chairs, kitchen tables and unpretentious upholstery. The stick-back or Windsor pattern chair, once made everywhere in Britain from various sorts of wood, became exclusively the product of High Wycombe – and it still is. The legs and sticks are no longer turned from cleft pieces by chair bodgers in the beechwoods; but the chair is still of beechwood, set into a seat of elm with a back frame of bent ashwood.

Imported birch mostly replaced beech for the interior frames of upholstered work, but beech was used for the best. It has now sometimes been replaced by English elm, because of the resistance of the latter to splitting under modern methods of pinning.

The demands of the furniture industry soon outstripped the supplies available locally from the beechwoods of the Chilterns and the elms of the Vale of Aylesbury. The timber yards supplied elm to the furniture industry in the forms of planks of about 1½ inches, and 'scantlings' (beams of small cross-section).

Besides the constant flow of coffin boards, the sawmills produced boards for tinplate boxes, blanks for pulley blocks, planks for bellows makers (a considerable industry in the Midlands), waggon scantlings, heavy planks for barge bottoms, boards for ammunition boxes, stocks and large felloes for wheelwrights and elm timber for

Indian clubs, dumb bells, croquet mallets and balls, bowls, skittles, etc.

Garden furniture, pit props and the cheaper clogs (usually alder) were often of elm.

The largest timbers were, and still are, supplied to the maintenance yards at docks and harbours where the massive baulks are used for all underwater work, including lock gates. The piles of the old Waterloo bridge were of elm and lasted 125 years, being still in good condition when taken up. The keels of wooden ships were usually of elm and various elm timbers are still used in boat building for keels and any parts remaining under water.

Arthur Ransome, whose boats were made to fit his dreams, describes in his autobiography the care of the craftsmen.

'I saw Selina's lines laid down on the floor of the loft. I watched the making of the moulds and Mr King's careful picking of the grown timbers that were to go into her. He and I went together to pick the elm for her yard-wide keel and saw the great tree lifted by a crane, laid on trolleys and driven in to meet the screaming saw that cut horizontally through the tree's entire length. Then the cut piece was lifted and we saw the top of the keel-to-be laid bare. You can never tell what may be in an elm until you cut it. The foreman of the timberyard was with us, and as the keel was bared and we saw the dark splash of bad wood in the middle of it, he said but one word, "coffins," and the tree that might but for that dark spot have sailed the seas was condemned to hold bones underground.

Two trees were cut through and condemned before we found the perfect keel.'

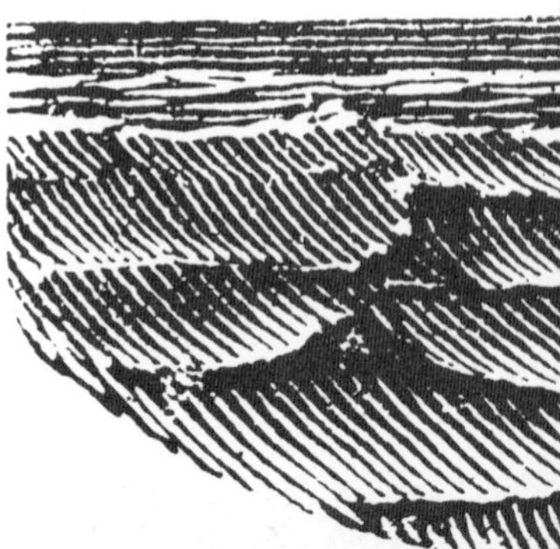

Elm lasts well under water but rots quickly at ground level, or, nautically, between wind and water. Timbers of canal engineering can be expected to be of elm. Wooden ships were of oak, but only elms were large enough for their keels: elm is still used for the keels of yachts and boats.

Hybrid elms, now ancient, at Hatfield Forest may have been specially grown for the angled timbers used by barge builders on the River Lea. The toughness of elm wood makes it ideal for sports equipment, from Indian clubs to croquet mallets, and the more modern skittles hockey sticks.

Elm Timber Today

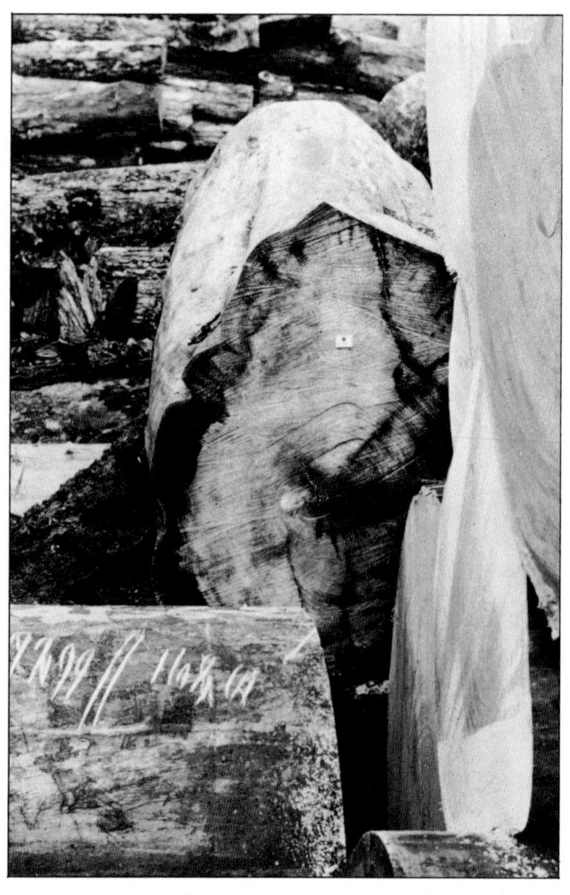

The modern timber converter requires his material to be even in texture and as near as possible to a perfect cylinder of unvarying length. Such a material lends itself equally to the diversities of the timber trade: baulks, planks, boards, squares, battens, scantlings, deals, strips, floor boards, match boarding, weather boards, skirtings, nosings, rebates, chamfers, and mouldings. It may be peeled for the veneers used for plywood and facings of blockboards, etc., and equally may be pulped for cellulose – everything from handkerchiefs to building boards. Pulping machinery, more than any other, requires logs of even size and quality.

The hedgerow elm does not conform to the standards set by forest-grown conifer wood. It is a hardwood, and the main demand is for softwoods. It is extremely variable, and though English elm forms a straight trunk, it rarely approaches a perfectly cylindrical form.

As a hardwood, however, elm in its various forms and varieties has its reputation in the timber trade and the demand for straight-grained elm in normal times equals or outstrips the supply.

English Elm (*U. procera*, various hybrids including Dutch elm and other English species)

Known also as the red elm, nave elm, European elm, Carpathian burl elm (in America, where its import is forbidden). Dutch elm was sometimes called sand elm.

The bole is usually clear up to 60 feet – the height of the mature tree being about 120 feet. The diameter of the bole is 3 to 5 feet, and up to 8 feet. Large trees are often rotten in the centre, though appearing sound.

The wood is reddish brown with yellow-white sapwood. The sapwood is considered to be as good as the heartwood. Elm is described as moderately heavy, firm, elastic, very tough, very difficult to split, extremely durable if kept dry or wet. But it is classified as 'non-durable', the test being how quickly it rots in contact with the ground. It tends to warp and twist, but rarely to split in drying. Its irregular growth and crossed grain produce 'an attractive figure' – the 'partridge-breast' pattern. Continental red elm is usually straighter grained. This is unlikely to be *U. procera*, which has no appreciable continental distribution.

Elm loses nearly half its weight in drying, reducing to about 34 lb per cubic foot compared with 45 lb for English oak, which weighs about the same when green: 65–67 lb. Shrinkage is even, being similar in both radial and tangential directions – that is why it does not split in drying.

Dutch elm is sometimes described as useless, being subject to heartshake, and sometimes as superior to and tougher than English elm.

The properties of English elm, lock elm, Dutch elm and so on are confused together. Some hybrids split more easily, some species are straighter

grained – there is no adequate literature on the subject. Timber merchants know their own districts, so do not need the information.

'This wood, although so well-known, is not treated with the consideration it deserves. There is little doubt but that scientific research could bring to light many more purposes for which it could be employed than those for which it is at present utilised' – so writes A. L. Howard in *Timbers of the World*, 1948.

English elm is sometimes difficult to work by machine, owing to its 'roey' and cross-grained character. It tends to bind on the saw, and to 'pick up' in planing. It is unsuitable for making plywood, but is sometimes used for decorative veneers. Two factories in Britain use elm chips for pulping to make brown corrugated paper.

The best timber now goes to the furniture trade, and there has been some revival of interest in elm coffins (which are nowadays usually made of

The grain of English elm

Peeled elm veneer is used as a packing material by brickmakers

chipboard). The problem for the Elm Marketing Board, an *ad hoc* body convened by the Forestry Commission, is one of a present surplus and a future shortage. You cannot stockpile the logs – they must be sawn within a year, or at most two years, nor can a tree killed by elm disease be left standing. An inquiry to the Council for Small Industries in Rural Areas, 35 Camp Road, London sw19, ought to result in an inspiring leaflet, on uses and sources of elm timber, and, one would hope, on possible markets for industrial components, containers, etc.

Considering the beauty of the wood I can only suggest that we should make and export elm furniture and kitchen fittings instead of importing 'teak' and pine ones – and look at those expensive wooden salad bowls. . . .

Old reference books refer to elm as a useful base or support for veneers. With modern cutting machinery and superlative glues it may now come into its own.

Wych elm (*U. glabra*)

Known as the mountain elm, Scottish elm, chair elm, *Bergrüster*.

The trunks of large trees can be 5 feet in diameter. The trunk often branches low but a clear bole up to 40 feet is not unusual. The timber is similar to red elm in appearance, but there is a clear division between sapwood and heartwood, and the heartwood may have a greenish tinge or green streaks.

The timber is generally considered to be superior to red elm, being straighter in the grain and less coarse. It may be as tough as ash, and it can be cleft; but is not prone to splitting. It is more elastic than red elm.

A special use of wych elm is in the seat planks of dinghies. It may be an adequate substitute for ash in chair making.

English elm and wych elm are used for veneers, forming about ten per cent of the work of a large veneer mill. Imported hardwoods such as sapele are the present basis of the trade, but English oak and yew are also cut into veneer wood to supply particular demands of furniture makers. Using massive and highly versatile machinery the logs are steamed, 'squared', and cut across, or quartered, or peeled. Quartering gives an approximation of a radial slice, with fairly regular veining. Such veneers, usually 0.02 inch thick, bedded on chipboard, can produce identical-looking runs of furniture in matching suites. Some manufacturers use the veneers in laminated blocks to get 'solid' elm without its tendency to warp and twist. It is perhaps typical of many industrial processes today that the best material should be selected, seasoned, steamed for hours, manoeuvred and jigged into position, sliced with incredible, automatic accuracy, dried in minutes, and then stuck together again. The whole of the cutting and drying can be done by a single large machine handling the log at one end and stacking the dried veneers at the other.

Thicker, peeled veneers of elm are supplied to brick makers as a substitute for straw in packing. The cores of peeled logs, about 10 inches in diameter, are sawn up and thrown into the boiler furnace.

English elm is justly popular with craftsman turners. Its resistance to splitting, its omnidirectional strength and its beautiful colour and grain make it suitable for bowls, plates and smaller pieces of tableware. Traditionally, nests of bowls were turned from a single piece on the pole lathe (which works by stops and starts in opposite directions). On the power lathe there is more waste, but elm bowls can compete with imported teak ones in price and quality. Mr Marchant, of Winchcombe, uses elm, including wych elm, for more than half his work. He makes salad bowls up to 4 feet in diameter, plates, and circular stools of his own design.

American white elm (*U. americana*)

Known as the water elm, soft elm, prinz wood; *l'orme parasol* or orhamwood in Canada.

It is usually 60–80 feet high but can reach 125 feet and a diameter of 7 feet. The timber is similar in weight to English elm, but is considered to be generally superior, especially when slow-grown and dense in texture. The sapwood is wide, of a very pale brown, and the heartwood is of a deeper brown, sometimes reddish. The grain is usually straight but can be interlocked.

White elm seasons easily and is very good for bent work, unlike the British elms, which tend to twist as well as bend. It is used for most of the same purposes. Ready-made wheel stocks and coffin boards used to be imported in competition with the home product. In Canada it is used in thin veneers for packing cheeses, etc., and in repairing casks. 'Quarter-sliced' (i.e. cut radially) American elm is especially decorative, a well-marked, parallel but wavy grain.

Rock elm (*U. thomasi*)

Known as Canadian rock elm in Britain, cork elm in Canada, cork-barked elm, hickory elm in the USA; cliff elm; sometimes white elm; and *l'orme à grappe*.

Rock elm reaches a maximum of 2½ feet in diameter; the stem may be clear for up to 45 feet. A straight-grained, light brown timber with no pronounced difference between sapwood and heartwood, it is the finest of elm woods in texture, the annual rings being finely marked; and it is the heaviest. It is one of the toughest timbers, 'seventy per cent better' than English ash and the equal of hickory. It is very resilient, straight grained but liable to split in drying. For this reason, when it is to be used in boat building, it is kept immersed in water or mud, and often kept in salt water for six months, being seasoned in a shed for a very short period. It is used for building wooden lifeboats, and in wooden vehicles and for parts of agricultural machinery. It was once used for the rims of bicycle wheels, from which it will be seen that it has excellent bending properties. It is not durable in contact with the ground or, nautically, between wind and water – like other elm wood.

American white elm of particularly dense quality is classed with rock elm.

The present virulent strain of Dutch elm disease is said to have entered this country in imported Canadian rock elm.

Slippery elm (*U. fulva*, syn. *U. rubra*)

Known as the moose elm, red elm, *l'orme gras*, grey elm or soft elm.

It has small sections, more durable timber than other elms and is more easily split, being used for fence rails, and for bending in boat building. Similar to American white elm in appearance, it is slightly heavier and often included with it. The newly cut wood smells of liquorice.

European white elm (*U. laevis*, syn. *U. effusa*)

Known as the spreading elm or *Flatterüster*.

It grows to 100 feet. The sapwood is broad and yellowish, the heart light brown. It is not clear how it may be distinguished from English elm, sometimes called European elm. Reputedly less strong than other elms, it is nevertheless valued for its markings for turnery and cabinet-making.

Chinese elm (*U. parvifolia*)

The timber is rich golden-red, with a close grain and no warp or twist.

Indian elm (*Holoptelea integrifolia*, syn. *U. integrifolia*)

It bears little resemblance to elm; the wood is light yellowish-grey to red, moderately hard and strong. It is used for door frames and cart building, and carving (according to Boulger, 1902).

Australian elm

Apananthe phillipinensis and *Duboisa myorporoides* are sometimes called elm in Australia.

False elm or **hackberry** (*C. occidentalis*)

This is heavy, not hard, but tough and takes a good polish. Sometimes it substitutes for elm.

Caucasian elm

The zelkova (*zelkova carpinifolia*)

The Elm in the Hedge

Almost all our elm timber come from isolated trees or small woods. The majority of elms grow in the hedges, of which there are 620 000 miles in Britain. The Forestry Commission census of 1951 gave the following totals, here in round figures, of usable timber (6 inches upwards) in hedges and parkland.

	cubic ft. (millions)	Some regional variations (% of hedge samples) Kent/Sussex	Hants/Cambs	Lincs/Norfolk
oak	250	50	16	17
elm	157	14	36	19
ash	105	32	40	43
others	296			

Some hedges consist entirely of elm. The Isle of Sheppey, in Kent, is notable for elm hedges, and part of this area is called Elmley Island. Ten per cent of hedges in Huntingdonshire, for example, are of elm.

The age of a hedge, with some exceptions, can be estimated by multiplying the number of tree and shrub species in a 30 yard length by 100 years. The system works whether the shrubs were added by man or nature. Relatively few hedges were planted originally with mixed species, while some others are sections of old cleared woodland and also started off with more than one species.

Some hedges are thus a sort of narrow scrubland, while others – the majority – are more like strip coppices, often including standard trees, under continuous management.

Standard trees in the hedge may be self-sown and allowed to grow up by the hedger, or they may have been planted with the hedge, or they may have resulted from later planting *in* the hedge – for instance, to fill a gap. Speculation about how a tree came into a hedge must take account of

a its distance from its neighbours in the hedge
b its performance as a colonizer of hedges
c its presence or absence in nearby woodland
d its popularity as a tree, for timber, for fuel, for decoration

Ash, the great pioneer tree in scrubland, is particularly quick to establish itself in the hedge, while hazel, historically a pioneer, is rare as a colonizer. It is typical of hedges in the oldest enclosed countryside.

Leaves collected in a few minutes from about three yards of hedge in East Suffolk

A rural elm avenue – for protection, ornament and use as timber

Thus a hedge may tell us something about the old dominant species of the area even though that area may be bereft of woodland. A predominantly elm hedge with a large number of subsidiary species will either be an old elm hedge or an old 'remnant' hedge invaded by elm.

The repeated presence of a particular elm clone in all the old hedges of a neighbourhood was used by Richens to plot the distribution of elms in Essex and other counties. But the gathering of such information requires minute and patient examination of dozens of leaf spectra,* and the possible results are of doubtful use.

Elm is a popular tree for hedges: the part of the hedge it shades out, it fills with its own suckers. Other trees, except ash, whose shade is light, tend to make gaps, which are often to be seen patched up with barbed wire. An unhedged boundary in arable fields, planted with a few elm trees, will become a hedge, as can often be seen.

Now that so many of the hedgerow elms are dying we should remember that their root suckers will remain and may be unaffected by disease for some years. The loss of the trees is very real, but it is unlikely to have either long or short-term serious effects on hedge ecology.

Pollard, Hooper and Moore, in *Hedges* (1974), list the birds which nest in the upper branches and

* Short shoots (not suckers or long terminal shoots) of elms usually contain about five leaves, described as a hand (see p. 49). Taking several examples of, say, the second leaf from a single variety or species, the botanist measures elements of mean leaf shape and arranges them in an arithmetical series, producing a 'leaf spectrum'. Where related spectra harmonize continuously with geographical distribution there is a strong argument for native origin.

White-letter hairstreak butterfly   Caterpillar of Vapourer moth

trunks of hedge trees: crows, rooks, jackdaws; mistle thrush; owls, stockdove, wood pigeon; blue tit, great tit; starling, tree sparrow and greenfinch, and the wren in ivy. The green woodpecker, the greater spotted woodpecker and the nuthatch are listed as less common. Many of these birds are of course woodland dwellers, only camping out in the hedgerow trees. They, and we, are fortunate in that woodland is never far away.

The elm is said to harbour up to eighty-two different insects, slightly more than hazel or ash, but against 149 for hawthorn, 266 for willow and 284 for oak.

Moore, Hooper and Davis (1967) selected hawthorn and elm hedges to survey relative bird populations. There were nearly twice as many bird species to be found in hawthorn: besides providing insects and berries, of course, hawthorn gives the most protection. The lowest yield of pairs of birds per 1000 yards was 2.3 from 'mechanically pollarded' elm. This is a type of hedge management we can expect to become general. The greatest numbers of pairs of birds per 1000 yards were 33.9 in bushy, neglected hawthorn hedges and in hawthorn-with-elm, 42.6.

The only creature entirely dependent on elms, including *Ulmus glabra*, is apparently the already rare white-letter hairstreak butterfly. The large tortoiseshell and perhaps the comma do feed on elm but not exclusively. The vapourer moth is said to haunt elm trees in London. Other moths sometimes found on elm are (culled from my Victorian moth book):

common quaker
chestnut moth
bright-lined brown-eye
light emerald
lime hawk
brindled beauty (has been known to strip whole trees)
early thorn
November moth

The loss of millions of trees from the hedge is a tragedy and a very sad sight, but it is not an ecological catastrophe – even the loss of considerable amounts of the hedges themselves cannot be called that, while we still have abundant small, or large woods. The dying of small woods of pure elm is a more serious matter, especially in hedge-bare districts, for the farmer is likely to remove the wood, rather than replace it by other trees. Here some amendment to the law on tree preservation ought to be made quickly.

The gradual conversion of lowland England to continuous highly productive field, segmented by continuous motorway, is said to be inevitable: yet, take a 'B' road, and you will almost certainly be reassured. The English countryside is complex, and the growth of wild plants is vigorous, even where every square inch seems to have been made productive. The elms will return in time. The death of elms, meanwhile, has made us aware of them not merely as the leafy frames for a picture of ripening cornfields and blue distances but as living roots.

Some Ancient Elms

On my desk as I write the last sections of this book, in cold, damp February, is a twig – full of hairy, round, pink buds. It comes from a tree which is now about 411 years old: it was planted, or germinated, around 1566, at a crossroads at 700 feet, and it was known as a landmark for two centuries. It is a poor old thing now – only a piece of the great shell remains, wrapped in bark as if expertly packed for a long journey, and a relatively junior branch rises crooked, like the remaining forefinger of a broken colossal statue. But even the fact that we know its age is the result of the affectionate care of the local people. Its top was blown down in a snowstorm in November 1893. The annual rings in a sawn section of the severed head were counted and the age of the tree was estimated with the help of the Forestry Commission.

This was, and is, the Sibford Cross Elm. The tree was adopted by Sibford School, which is run by the Society of Friends. A tree assumed to be its offspring, on the opposite side of the road, has now been invested with the honour, and a small corner of land at this windy, remote crossroads in the east Cotswolds now contains a couple of seats, for contemplating the elms.

The Sibford Elm was a landmark tree, visible for miles, and some other old elms on hilltops have been known by name, like the Nell Ball Elm at Plaistow Mount in Sussex. The list of oaks is much longer of course – elms are old at 200 years.

A habit of throwing out one strong side branch has made elms useful as gallows. One at Alderton in Gloucestershire is named Dabb's Elm after a sheep thief who was hanged there; and at Branch's Cross, Wrington, Somerset, an ancient elm is preserved from which were hanged Branch *and* Cross, in 1685. Rupert's Elm, at Henley, is where Prince Rupert hanged a Roundhead spy.

The Gossip's Tree at Fontmell Magna, Dorset, is a venerable pollarded elm formerly known as the Cross Tree. Preaching under elms perhaps requires more confidence than spreading the Word under oaks – the title of Gospel Oak is a familiar one. But Wesley conducted his open-air ministry from several now ancient elms. One, about 300 years old, remains at Stony Stratford (on Watling

Rupert's elm at Henley. Above right and opposite **The Sibford Cross Elm— the new tree and the old**

Street, Bedfordshire) and another, a Cornish elm, is near Camborne in Cornwall. A 'massive elm' under which Wesley preached was at Hilsea, Portsmouth, and is now replaced by a young tree with a plaque.

Great elms have become famous for being hollow. Smugglers hid their booty in three elms called the Old Maids, near Rochford, Essex. A hollow tree called the Crawley Elm, on the Brighton Road, was 61 feet in circumference in 1838, and 35 feet on the inside. The floor was paved with bricks, and in the trunk there was a door to which the lord of the manor held the key. Parties were held in this elm, and a poor woman once lived there and gave birth to a child.

In the seventeenth century there was another famous hollow elm at Hampstead. A staircase inside led up to a platform with seats for six people to enjoy the view. The engraving by Hollar probably shows exactly what it was like. Also in London, and still there, is the oldest elm in Britain – a living trunk 12 feet high, held together with chains. It is in Half Moon Lane, SE15. Wilks says that it was well known in Chaucer's time. The Seven Sisters are elms at Tottenham, commemorated in the name of a long road in north London. They were first planted by the seven sisters of Robert the Bruce, then replanted in 1886 by seven other sisters called Hibbert. Nine Elms is an old part of Battersea now occupied by a famous power station.

The village maidens went, in May, to dance round the Tubney Elm, between Fyfield and Abingdon, as recorded in Matthew Arnold's *Scholar Gipsy*. This tree was 37 feet round, and hollow. A stump still remaining, but dead, at Guarlford, Worcestershire, is the Friar's Elm, where a priest once distributed bread to the poor.

Two elms are recorded as having second sight: the Prophet Elm at Greenhill Court, Herefordshire, shed a branch whenever a member of the family was about to die. A prophetic elm at Hinton-in-the-Green, Worcestershire, had the curious reputation of signalling a drop in the male birthrate, should it fall down. It fell, but I do not know if there are too many girls today in Hinton-in-the-Green.

All these and several other elms are mentioned by J. H. Wilks in *Trees of the British Isles in History and Legend*, 1972. The author gives also the locations of fine specimen trees. We note

U. procera	122 foot high, nearly 20 foot round (at 5 foot). Youngsbury, Ware, Hertfordshire.
U. carpinifolia	118 foot. Kensington Gardens. 85 foot nearly 20 foot round. Dane End, Hertfordshire.
U. × hollandica	111 foot. Saltram House, Devon. Var. *vegeta* 90 foot, 17 foot 4 inches round. Westonbirt village, Gloucestershire.
U. angustifolia	Var. *cornubensis* 118 foot. Knightshayes, Devon. *sarniensis* Kew Gardens, north-east corner.

and a mighty wych elm, 125 foot, at West Dean, Sussex.

GATED
ROAD
WIGGINTON
SHUTFORD

Part 5
Elm Disease

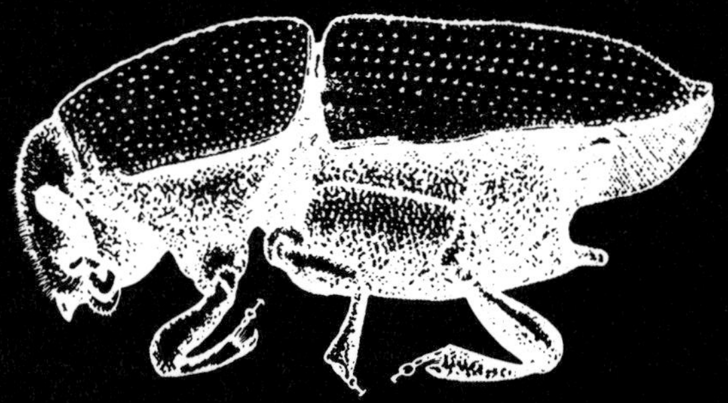

Elm Disease

Winter 1976

Summer 1976

Some country people who should know better seem to think that every tree with Dutch elm disease is a Dutch elm. The name is unfair to the Dutch, who identified but did not nurture the disease, discovered in France in 1818, and in fact have contributed much in attempts to develop disease-resistant varieties. The disease is a fungus, perhaps of Asian origin. It was known to Loudon in 1838; he reports a solemn and unscientific experiment intended to prove that the bark-beetle which carries the disease fed only on rotten wood.

The strain found in 1818 and identified in Holland in 1919 began to be noticed in Britain in 1927. Surveys begun in 1928 showed that the disease was already widespread. It began to decline in 1937. A lot of trees recovered, being only partly infected, and the disease remained at a low level until the present – it is described as non-aggressive and rarely fatal. This epidemic killed between ten and twenty per cent of all English elms.

A new, much more virulent form of Dutch elm disease came in the 1960s, killing an annual 400 000 trees in America and Canada. It reached Britain before 1970, and by 1974 had killed 4.5 million of our elms. Over fifty per cent (732 000 over 20 feet high) of the elms in Gloucestershire were dead or dying in 1973,* and this figure will probably apply to the whole of southern England by the time this book appears:,approximately one-quarter of the trees will be classifiable as good timber. They will sell at less than a third of the price of English oak and being valued low are unlikely to be stored against the inevitable future shortage. The effect on the landscape is, of course, miserable.†

Surveys conducted for the Forestry Commission by J. N. Gibbs and R. S. Howell show the astonishing rate at which the epidemic has taken hold, though the figures are somewhat inaccessible within a twiggy growth of estimates and subcategories. By the end of 1971, 600 000 trees out of an estimated 18 000 000 had been killed (3.3 per cent). In 1972 the total elm population was estimated at 22.2 million, the growth being attributed partly to 'observer error' and partly to the number of trees reaching countable height (6 metres, about 20 feet). Of these, 4.3 per cent were dead or dying (i.e. more than half the crown affected). By 1973, 12.8 per cent were dead or dying, not including those long dead (more than two years). The surveys have considerably clarified the distribution patterns in south England of the three major taxonomical groups: *Ulmus glabra*, *U. procera* and *U. minor*. Confusingly *U. carpinifolia*, with *angustifolia*, *plotii* and, no doubt, *coritana*, are included under *U. minor* in the surveys, while hybrids take their chance in any group; the silent

* This figure includes elms long dead but still standing.

† Eleven million elms are now lost, according to a Forestry Commission report published in November 1977 (see pages 144–45).

This chart is based on the Forestry Commission report on elm disease for 1976, the last year in which individual counties were surveyed. The disastrous effect of the disease on Britain's elm population can be seen from the figures returned for that year. Those counties which were included in the survey are grouped under general tables.

Each area table shows: the total number of trees sampled (all tree symbols) those affected by the disease, and the remaining healthy trees. The symbols represent all species of elm.

The smaller diagrams show the decline by county of healthy elms since 1972 according to Forestry Commission figures.

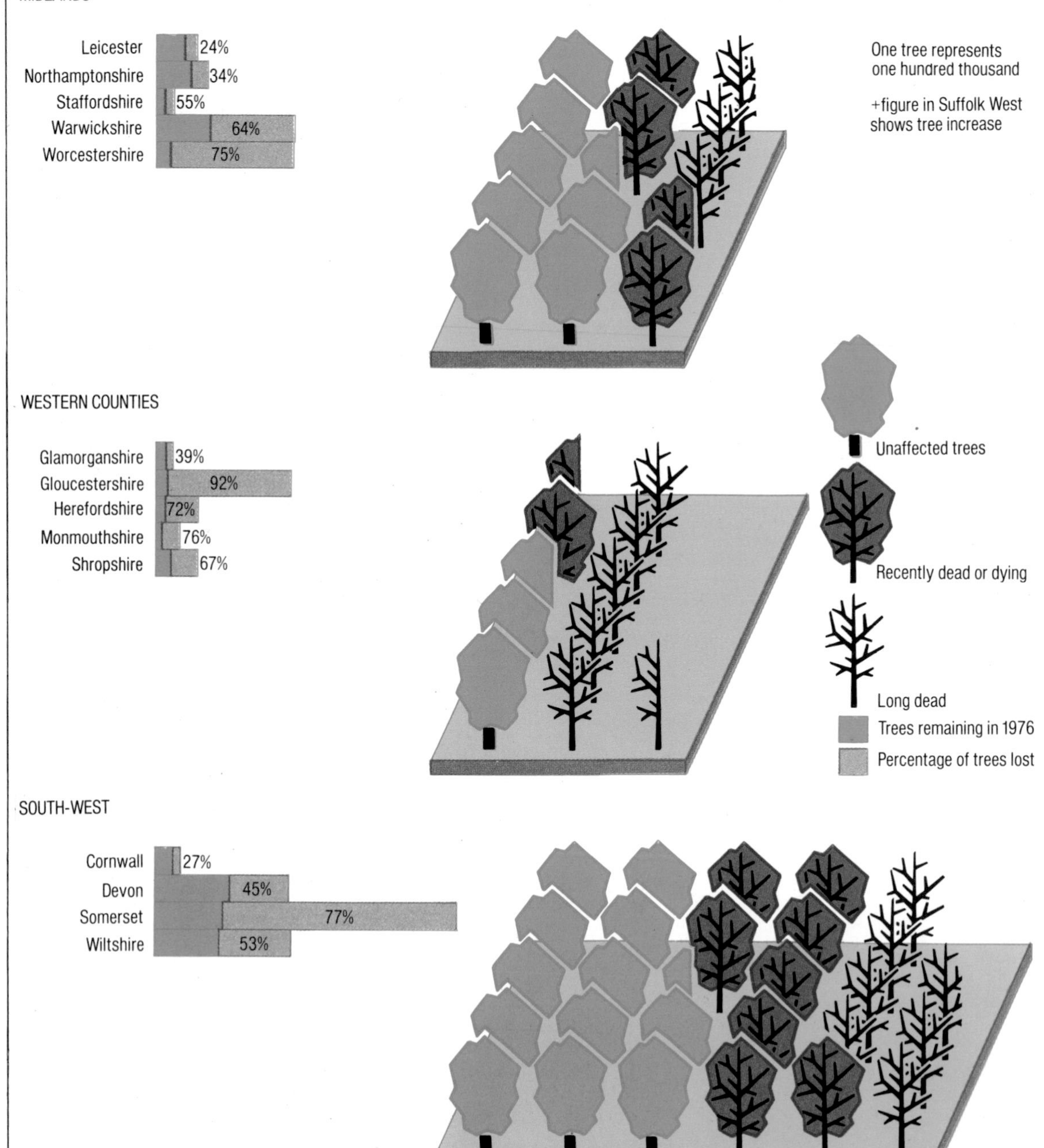

Any attempt to make an accurate assessment of trees killed by Dutch elm disease over a period of time is difficult because the decrease in total trees recorded from year to year is partly caused by trees being felled or cleared. Also, estimates of infected trees in individual counties are subject to sampling error. However, it is safe to assume that the number of healthy elms in 1971 was markedly greater than the total number of all elms shown here.
A Forestry Commission report published in November 1977 estimated that eleven million elms have been killed by Dutch elm disease and that in some areas, such as West Midlands, Surrey, Hampshire and West Sussex, most elms have now been lost. The worst affected areas are, according to the report, south of a line from the Mersey to the Wash but even outside these areas there has been a significant increase in the spread of the disease over the past year: the number of reported cases has been two to five times higher than during 1976 in northern England and western Wales.

HOME COUNTIES

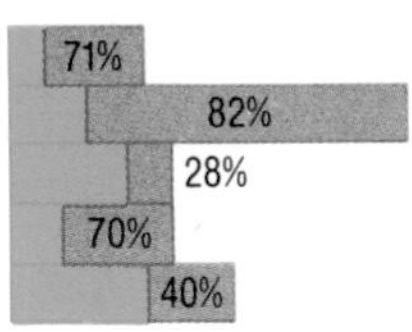

EASTERN COUNTIES

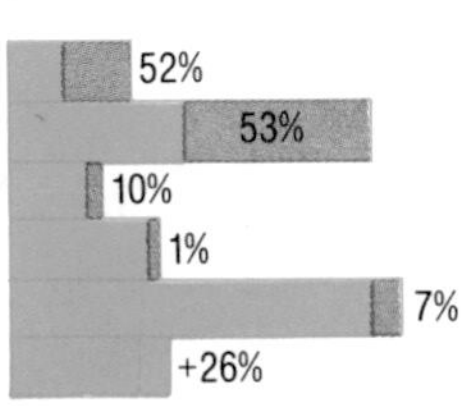

SOUTH AND SOUTH-EAST

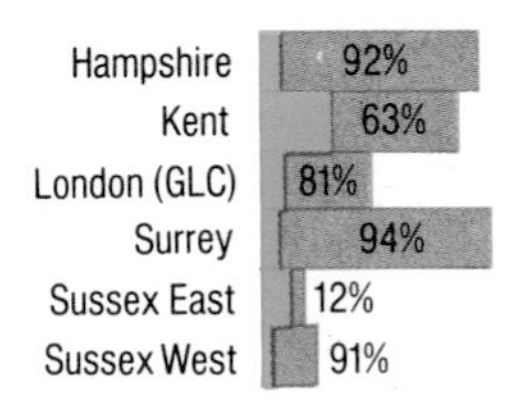

majority, and just as dead as the others. *U. minor* (any smooth-leaved elm) has acquired a slight reputation for ability to recover.

Gibbs and Howell also sampled the northern counties' elms by selecting at random fifteen 10-km squares from likely elm country; seven-eighths of the elms were wych elms. Disease was little in evidence.

Dutch elm disease has in fact affected most severely those areas with heavy elm populations. Neither the wych elm nor the smooth-leaved group are usually found in heavy concentration and part of their supposed reputation for resistance must result from the manner of their distribution.

Concentrations of high incidence of disease in the Severn–Avon valleys, in South Hampshire and south-west Sussex, and in Essex, are near the major ports, and it was this pattern which indicated to Brazier and Gibbs (*Nature*, vol. 242, 1973) that the present vigorous strain enters this country in imported wood. They identified the strain (and found live North American beetles) in logs of Canadian rock elm (*U. thomasii*) from Toronto. It appears that the new strain originated in Europe and spread to America before returning across the Atlantic. Imports of European elm into the USA are forbidden.

All elms are affected, including the Chinese elm, *U. parvifolia*, once thought to be immune. Hybrids of *U. parvifolia* with *U. americana* and others were planted in extensive experiments during the epidemic by the US National Arboretum in an attempt to find a resistant hybrid. *U. americana* itself is highly susceptible. *U. pumila* is said to show remarkable resistance, though not consistently. *U. villosa*, which looks even less elm-like, one might expect to be immune – but other species have proved disappointing before. Gibbs and Howell stated in 1974 (*Forest Record*, no. 100) that there was some evidence for greater resistance in *U. glabra* and *U. minor*, and that Huntingdon elms were rarely affected.

While resistance to the disease may vary slightly with species, it does seem that habitat and area of distribution are the governing factors in rates of infection. From my own observation I am tempted to theorize that elms remote from busy roads have their beetle populations controlled by bark-feeding birds, but no one working in the field has mentioned this. Certainly one could collect many examples of rows of elms at right angles to roads where the first trees to be infected are the ones

Ulmus carpinifolia, the crown half infected

nearest to the road. But there are also many examples of isolated, remote trees dying of the disease.

Cornish elms in Cornwall are only locally affected. Bedfordshire, Cambridgeshire, Hertfordshire, Huntingdonshire, Suffolk and Norfolk, with predominantly smooth-leaved and hybrid elms, appeared in 1973 to be less severely hit than the south Midlands – the rate of increase, after the initial encroachment, was less. But Hertfordshire is now one of the worst affected counties. West Essex, Hampshire, Warwickshire, Worcestershire, with large *U. procera* populations, were severely affected. The highest rates of increase between 1972 and 1973 came in counties which had been relatively little affected previously, says *Forest Record*, no. 100: Oxfordshire, Wiltshire and Northamptonshire. West Sussex, with eighty-eight per cent dead elms in 1976, differs remarkably from East Sussex, with only ten per cent. There may be a concealed contrast in elm habitat in the two halves of the county, but the numbers of elms are about the same. East Sussex County Council attributes disease-free conditions of its territory to the vigorous programme of felling and of publicity conducted by Tony Denyer, elm disease officer for East Sussex. Perhaps he was lucky to have a fairly environment-conscious body of citizens, able to spend about £40 on disposing of a diseased tree.*Analysis might also reveal a tendency for East Sussex elms to be close to or in parks, gardens and dormitory villages – this is not to denigrate the success of the campaign and the energy with which it was pursued at all levels.

The national policy of 'sanitation felling' (under

* This high average takes into account the difficulty of access to smaller plots, but does not allow for the value of the fuel, let alone the timber. For a whole park-full of Huntingdon elms in north London, determined residents with an option to buy from the Council could not find enough profit from the sale of the lumber to cover the cost of its transport to the nearest merchant.

The cost of removing all the dead elms of Britain has been estimated as that of 1000 men working for five years, or £7 million per year – that is, twice the men's pay. But if only one quarter of the trees yielded an average 135 cubic feet of timber at 30p, the operation would show a profit of £56m before sawing and storage. Lop and top at about $1\frac{1}{2}$ tons per tree would yield another £150 million (delivered to user, if users could be found). In fact most of the trees rejected for timber could still provide excellent wood if handled by specialised plant.

A spin-off would be an appreciable reduction in the spread of the disease, while the improvement to the countryside would at least raise our spirits. But of course the enterprise is unlikely. All the dead trees belong to somebody, and the owners have to choose between paying to have them taken away, salvaging them for firewood, or leaving them to fall and rot.

Scolytus beetle taking off after drying out on the stump of its host tree, and below a feeding wound

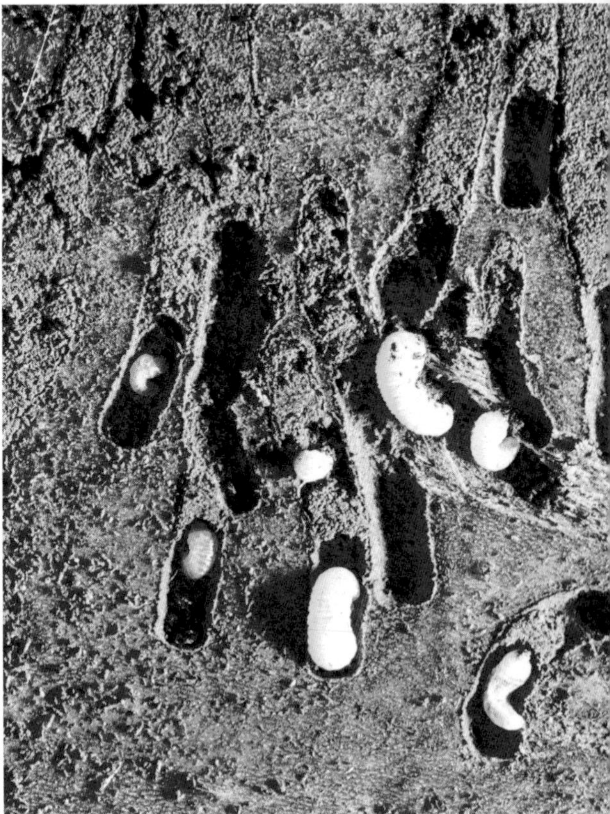

The larvae of the scolytus beetle in their galleries, which are partly blocked with their excreta, under the bark

the 1971 Dutch Elm Disease Local Felling Order) was dropped in 1972, in spite of East Sussex and in spite of reports from America that ring felling – isolating diseased groups – had kept the disease down in some areas. Illinois, with below one per cent diseased trees in twenty-six cities, was quoted: lopping of infected branches and injection of fungicide had kept the losses of town trees down. Local authorities in Britain still have powers of inspection and control where felling still offers some prospect of controlling the disease. A local authority may serve a notice on a responsible person requiring the destruction of diseased trees; failure to comply is a penal offence. An Order of 1974 prohibits the movement of diseased elm wood in parts of Britain where disease levels are low. (Details, including the scheduled area where this Order does not apply, are obtainable from the Forestry Commission in London.)

But the policy is, broadly, to let the disease take its course. Individual trees of 'high amenity value' (i.e. somebody loves them) may be treated chemically by the methods described below.

The disease is caused by a fungus *Ceratocystis ulmi* (*Ceratostomella ulmi* or *Ophiostomata ulmi*) the non-aggressive and the very pathogenic forms of which exist side by side – there are said to be visible differences. The fungus is like a yeast carried in the sap of the tree. It contains a toxic substance which stimulates the walls of the sap-carrying vessels to produce bulbous, gum-filled enlargements called tyloses, which are usually found in the (dead) heartwood of trees. The vessels, or tracheids, are thus blocked and the flow of nutrients is stopped. A section across a diseased twig or branch may show a ring of dark brown spots in the sap wood – these show as streaks if the bark is removed. The blocking of the sap vessels sometimes causes a resin-like substance to be exuded from the bark.

The first signs of infection on the tree are drooping or curling shoots and yellowing leaves in summer. Usually one branch only is affected – it may die in a few weeks. The curled twigs persist through the winter. The fungus is usually brought to the tree by a bark-beetle: *Scolytus scolytus* (*S. destruc-*

Nuptial chamber, main channel and radiating feeding galleries

Trees in the first stage of infection

tor), the large European elm bark-beetle, or sometimes by the much smaller *S. multistriatus*.

These beetles breed under the bark of recently dead elm trees. The female tunnels into the bark as far as the sapwood, and is joined by the male who prepares a nuptial chamber. After mating, the female burrows along the grain between the bark and the wood, laying a string of about seventy round, white eggs, alternately in the left and right walls of the tunnel. The legless, white larvae, when they emerge, begin at once to feed and tunnel at right angles to the main gallery. Their channels are less than 0.02 inch wide and about 0.04 inch apart. The grubs grow to nearly ten times their original size, changing skins five times, over a distance of about 5 inches. The tunnels get gradually wider and fan outwards, never joining or crossing but sometimes tangling with the galleries of neighbouring families. The larvae make slightly enlarged chambers to pupate.

In May to October, during warm spells, the young adult beetles emerge through circular holes. They fly away, perhaps several miles, to feed on sap, mainly in the tops of the elm trees. On finding a suitable host a beetle gives off a scent to attract its fellows. They make incisions usually in the crotches of healthy four-year-old twigs. Heavy attacks can kill the twigs by ringing the bark, but the beetle invariably carries the spores of *Ceratocystis ulmi* both externally and internally: a future breeding site is assured.

Bark-beetles, of which there are sixty-six species in Britain, are often called engravers, because of the distinctive linear pattern they leave on the inside of the bark, appearing as if printed in reverse on the wood. Other names are shothole borers, referring to the egress holes, and ambrosia beetles, because of their association with the fungus, thought to be cultivated by their excrement and to provide their food. Whether the fungus supplies essential food elements to the grub is not clear, but the aggressive strain of *C. ulmi* can certainly spread without the aid of the beetle, at least between the roots of adjacent trees, while timber merchants can take the place of the beetle in its wider distribution, even from one continent to another. The

bark of dead trees and branches may attract the beetles for two years: after that it is ignored.

The fungus was described by Gibbs in 1974 (*Forest Record*, no. 94). Its fructifications are of two main kinds. The first, the imperfect stage (*Graphium ulmi*) is asexual and consists of thin black stalks 0.04 inch (1 mm) high with sticky heads containing the microscopic spores. The second is sexual in origin; black flask-shaped bodies with long narrow necks. The spores ooze out of the bristle-like neck forming a gluey drop at the end. This fruit body is only 0.02 inch (0.5 mm) tall. Both types are visible in quantity to the naked eye, but hard to find: they form in the beetle galleries.

Other funguses may cause wilting, yellowing of the leaves and die-back: *Verticillium*, *Phytophthora* and *Nectria cinnabarina*, or coral spot, the bean-pole fungus. Honey fungus (*Armillaria mellea*) may kill trees – so may gas-main leaks. Leaf-curl and vein galls in the leaves are caused by aphids, and two species of moth make mines in the leaves, which are also eaten by various caterpillars, notably the white-letter hairstreak and some Vanessas. Brown spots on the leaves are caused by a mite. Wetwood disease is bacterial.

Bark-beetles of all trees, usually carrying fungal diseases, are normally kept in check by birds which pick out the larvae. Some ambrosia beetles are suffocated by luxurious growth of their associated fungus. Parasitic wasps and predatory larvae of other insects including other beetles also play their part in keeping the balance.

In a Colorado spruce forest, where woodpeckers were normally seen to increase following any increase in beetle life, a great storm blew down an unusual number of trees. The woodpeckers could not feed on the fallen trunks, and a bark beetle which normally had been beneficial in thinning the forest of senile trees increased astronomically over a period of three years, until swarms of beetles attacked healthy trees, bringing also a deadly fungus. Six years later millions of trees had been destroyed, many times more than the considerable total lost by fire over thirty years.

This forest disaster was caused by one severe

storm. The story shows how one relatively small ecological change can lead to enormous destruction.

The present epidemic is only one instance of the sensitivity of the natural environment to the massive waywardness of twentieth-century technology: in this case our facility of transport. While we carry pathogenic strains of tree fungus about the world, unchecked, do we have an equal technology to combat the effects when they get out of hand? It seems not.

Even the current wave of elm death would not be outside the scope of natural predators. Cuckoos are said to eat more than 1500 Scolytidae a day – it sounds like hard going. Robins, tits and nightjars eat insects, and starlings, as well as the trunk-dwelling woodpeckers and tree-creepers, eat grubs out of tree bark. We would need a phenomenal expansion of these bird populations perhaps; but even the less-shy birds would avoid the trees in the third of our hedges that border roads. The 'quiet' country lanes are all busy now: transport again. And our fossil-fuel economy despises lop and top, deadwood and old tree roots – they do not go with thermostatically controlled central heating, or with coal at a guinea a bag fed into tiny grates. The countryside was full of unsightly dead trees and rotting branches before anyone began to worry about diseases. Beeches and sycamores are the next to be threatened, then the native ash. During a dry spell fungal spores are very easily dispersed. The unusual frequency of fungal diseases in trees in 1977 may be one result of the extreme drought of 1976, and not a sinister trend.

Treatment

The only sure treatment for *Ceratocystis ulmi* is felling the tree and burning the bark. The timber is not damaged. As an alternative to burning, the bark may be treated with an insecticide called Lindane. This chemical should be diluted with kerosine for winter treatment – water will do after March, as the insecticide will probably persist until the summer flight of the beetles. Already infested, stacked logs cannot be successfully treated by this method.

To preserve the life of a tree, pruning an infected branch close to the trunk is recommended. A tree should not be felled at the first sign of disease: it may recover.

However, the fungus may spread to an adjacent tree up to 30 feet away through the roots. English elms send out long suckers, and their roots also are apt to graft together with brother trees; if they were not in fact from the same root to begin with, or are not parent and offspring. Trenching machines which can dig down to 2 feet can be used to sever these underground links – if there are no pipes and conduits in the way. Chemical treatment can also be used to kill the roots along a narrow band between the trees. A soil sterilant, metham sodium, diluted 1:4 with water, is poured down holes 2 feet deep, 6 inches apart, made with a soil auger. Trees within 6 feet (2 m) of the drills will be in danger, and all vegetation, including grass, will die along a foot-wide strip. The infected tree which this treatment is intended to isolate should not be felled until two weeks after treatment, as this would stimulate the fungus in the roots. Secondary barriers are recommended, separating healthy trees beyond the apparent zone of infection.

Fungicide injection

A laborious method has been evolved which can be used to save the life of a particularly valued elm. Carbendazim hydrochloride, manufactured as Lignasan, TBZ or M & B 21914, or Benomyl, is injected into the sap stream of the tree, at a convenient height on the trunk. See *Forestry*, vol. 48, no. 2 or apply to the Pathology Branch of the Forestry Commission. The treatment has to be repeated annually, but it does work. Sadly, the keepers of our royal parks do not seem to have been able to afford it.

Spraying with insecticides

This method uses various types of mechanical mist-blowers to drench the whole tree, especially if possible the upper branches, with a chemical intended to poison the young adult beetles as soon as they begin to feed in early summer. The insecticide must be applied before the leaves open if it is to reach the twigs where the feeding wounds are made; but as late as possible because its protection lasts only a few weeks. The most effective chemical is methoxychlor (Marlate) in emulsifiable concentrate (e.c.) or wettable powder (w.p.). DDT was formerly used, but abandoned for this and other outdoor purposes because it builds up poison in feeding animals and birds. Methoxychlor is lethal to fish.

Spraying from a helicopter covers the tops well

Injecting an elm tree with fungicide, plugging the hole and painting over

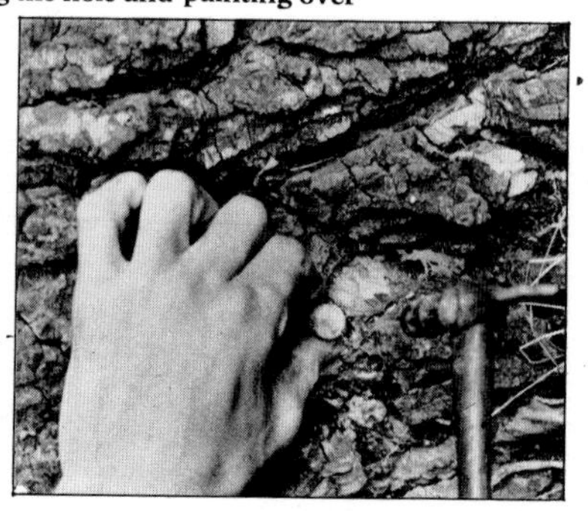

but does not reach the lower branches. From the ground, a Coventry Climax fire pump overcame the effect of wind but had a very high output, difficult to control. For town trees, low volume sprayers can be used with knapsack mistblowers from hydraulic platforms.

The insect must feed before it absorbs the poison, and in theory only one wound is needed to infect the tree. See *Forest Record*, no. 105, for a full account of these experiments.

Austrian wasps whose larvae are predators of *Scolytus* larvae were imported by Basildon New Town, apparently without any success. Even so, biological control is the only solution to the problem, even if it must take the form of a shortage of food/host plants. Young suckers, from trees living or dead, seem to be immune for six to ten years. But I have found feeding wounds on very young trees. The bark would not be suitable for breeding, but the infection may remain.

Part of a pamphlet produced by East Sussex County Council

Replacing The Elms

The native black poplar

Clearly one positive response would be a vigorous planting campaign. A low-priced Forestry Commission Leaflet (no. 57) by A. F. Mitchell recommends the following: common oaks, Turkey oak, beech, sweet chestnut, small-leaved lime, large-leaved lime, ash, grey poplar, sycamore, hornbeam, white willow, Italian alder, Norway maple, common alder, crack willow, grey alder, gean, bird-cherry, field maple and wild service tree. The author does not mention the black poplar, a noble and native tree which has some of the character of elm in its massive blistery trunk. To replace tall trees, Canadian poplars also can be impressive.

The cost of planting a tree is about £5, if you buy the tree and employ labour to plant and fence it. However, willows and poplars can be grown for free – almost any live stick will strike root. Sycamores sow themselves nearly everywhere in the western half of the British Isles, and some hopeful work can be done by scattering acorns and maple seeds in likely-looking places. Young ash trees can be removed surreptitiously from hedges where they are crowded to places where they are needed. Try setting wild service berries (chequers) in pots or seed beds, for planting in odd corners. There is a good tree on the north-east side of the Hampstead Heath Extension (map reference 26.88). Avoid alien species, except for the poplars: study the surroundings. Local landscape gardeners may know of impending clearings which could provide instant shrubs and saplings for the trouble of digging them up (with plenty of soil attached).

The crying need is for miniature nature reserves of a completely unpretentious and unpublicised sort. They only need rabbit fences and immunity from tipping: nature will do the work, at least until you have defined the pattern of management. Who provides the land? Be bold, and ask the local council, landowner or farmer. You might even buy a bit of land with no building permission: or borrow a bit of garden.

The Forestry Commission can now provide grants for planting anything down to 0.6 acre (0.25 hectares) and the Countryside Commission can help with areas less than that.

If you own parkland and want to plant an avenue – well, last year I would have said, plant Huntingdon elms and pray for them. But revisiting Hidcote in 1977 I have lost confidence even in this most vigorous of the elms.

Book List

'Acorn', *English Timber and its Economical Conversion*, Rider, 1910.

J. Aubrey, (ed. Britton) *Natural History of Wiltshire*, 1848.

J. Betjeman, *Collected Poems*, J. Murray, 1972, and *Summoned by Bells*, J. Murray, 1960

G. S. Boulger, *Wood*, Arnold, 1902.

J. C. S. Brough, (ed.), *Timbers for Woodwork*, Evans, 1947, revised 1962.

A. R. Clapham, *Flora of the British Isles*, Cambridge University Press, 1962.

J. Clare, *Poems*, Dent, 1974.

H. L. Edlin, *Woodland Crafts in Britain*, David and Charles, 1974.

H. L. Edlin, *Trees, Woods and Man*, Collins, 1970.

H. J. Elwes and A. Henry, *Trees of Great Britain and Ireland*, 7 volumes 1906–1913.

T. H. Everett, *Living Trees of the World*, Thames and Hudson, 1969.

H. E. Godwin, *History of the British Flora*, Cambridge University Press, 1975.

M. Grieve, *A Modern Herbal*, Cape, 1974

J. Gerard, *The John Gerard Herbal*, Dover Publications.

M. Hadfield, *British Trees*, Dent, 1957.

HMSO, *Handbook of Hardwoods*, 1973.

H. A. Hyde, *Welsh Timber Trees*, National Museum, Wales, 1961.

B. Langley, *Landed Gentleman's Companion*, 1741.

C. R. Leslie, *Memoirs of the Life of John Constable*, Phaidon, 1971.

J. C. Loudon, *Arboretum et Fruticetum Brittanicum*, (8 volumes, see volume 7), 1838.

F. K. Makins, *The Identification of Trees and Shrubs*, Dent, 1967.

A. Mitchell, *Field guide to the Trees of Britain and Northern Europe*, Collins, 1974.

W. Pennington, *History of British Vegetation*, English Universities Press, 1974.

F. H. Perring and Walters, *Atlas of the British Flora*, Nelson, 1968.

E. Pollard, Hooper and Moore, *Hedges*, Collins, 1974.

H. Prince, *Parks in England*, Pinhorn, 1967.

O. Rackham, *Trees and Woodland in the British Landscape*, Dent, 1976.

S. Ross-Craig, *Drawings of British Plants*, part XXVII, Bell, 1970.

E. Thomas, *Collected Poems*, Faber, 1949.

J. H. Wilkes, *Trees of the British Isles in History and Legend*, Muller, 1972.

Forestry Commission Leaflets

D. A. Burdekin and J. N. Gibbs, *54 Control of Dutch Elm Disease*, 1974.

A. F. Mitchell, *57 Replacement of Elm in the Countryside*, 1973

Forest Records

J. N. Gibbs, *94 Biology of Dutch Elm Disease*, 1974.

J. N. Gibbs and R. S. Howell, *100 Dutch Elm Disease Surrey 1972–73*, 1974.

T. M. Scott and C. Walker, *105 Experiments with Insecticides . . . Dutch Elm Disease*, 1975.

Articles

R. Melville, Goodyer's Elm (1938), The Plot Elm (1940), Ulmus Carpinifolia (1946), The Coritanian Elm (1949), *Journal of the Linnean Society (Botany)*; The British Elms, *New Naturalist*, 1948.

Acknowledgements

The author wishes to thank Sir John Betjeman and John Murray Ltd for permission to quote from *Collected Poems* and *Summoned by Bells*; Faber & Faber Ltd for permission to quote from poems by Edward Thomas; and Jonathan Cape Ltd for permission to quote from Ted Walker's *Fox on a Barn Door*. The information about the Sibford Elm comes from Leslie Baily's collection, University of Leeds, and for the disease chart on pages 144–45 from the Forestry Commission. Thanks also to Rory Shepherd, of Mallinsons Veneer Mill, Lydney, for his time and interest, and Christina Fox for typing and help with research.

Credits

Photographers

A–Z Collection 64, 149
Heather Angel Ltd 43, 47 (top), 89 (main picture), 95, 109, 148
Mark Edwards 155
Ercol Funiture Ltd 5, 129
Tony Evans 13, 25 (top left)
Forestry Commission 55 (2), 152 (second from bottom)
Hazel Geoffrey 57, 72, (top right)
Fay Godwin 44, 48, 143, 150/151
John Hillelson Agency (Dr Georg Gerster) 112
Bernard Higton 16/17, 19, 20/21, 24, 25 (bottom left), 110, 117, 127 (top left), 152 (top), 152/153
Mid West Photographic Agency 26/27
Stockphotos International 85
Gerald Wilkinson 3, 8/9, 14/15, 18, 22, 23, 25 (top and bottom right), 44, 47 (lower), 49, 51, 58, 59, 60/61, 62/63, 65, 66/67, 69, 70, 71, 72 (top left), 73, 74, 75, 76, 77, 78/79, 80, 81, 82, 83, 84, 86, 87, 88, 89, 92/93, 95 (inset), 96/97, 99, 104/105, 106, 108, 116/117, 126, 127 (top centre), 130, 132, 133. 137, 138, 142, 146/147, 148 (left), 149 (right), 152 (second down and bottom), 156, 157, 160

Artists and illustrators

The Diagram Group 54/55 (trees), 100/101, 144/145
Graham Evernden 120/121, 122/123, 124
Peter Knock 38/39
Ben Perkins 50, 56, 67
Gerald Wilkinson 54/55 (leaves)

Museums, Art Galleries and Picture Collections

British Museum, prints and drawings 136/137
Mary Evans Picture Library 127 (top right)
Glasgow Art Gallery 30 (lower)
National Gallery 28
Godfrey New 116
The Science Museum 125
Tate Gallery 33, 34 (photo John Webb)
Walter Morrison Picture Settlement Trust, 33